Women and Wildlife Trafficking

This volume examines women and wildlife trafficking via a collection of narratives, case studies and theoretical syntheses from diverse voices and disciplines. Wildlife trafficking has been documented in over 120 countries around the world. While species extinction and animal abuse are major problems, wildlife trafficking is also associated with corruption, national insecurity, spread of zoonotic disease, undercutting sustainable development investments and erosion of cultural resources, among others. The role of women in wildlife trafficking has remained woefully under-addressed, with scientists and policymakers failing to consider the important causes and consequences of the gendered dimensions of wildlife trafficking. Although the roles of women in wildlife trafficking are mostly unknown, they are not unknowable. This volume helps fill a lacuna by examining the roles and experiences of women with case studies drawn from across the world, including Mexico, Cameroon, the Central African Republic, South Africa and Norway. Women can be wildlife trafficking preventors, perpetrators and pawns; their roles in facilitating wildlife trafficking are considered from both a supply and a demand viewpoint. The first half of the book assesses the range of science, offering four different perspectives on how women and wildlife trafficking can be studied or evaluated. The second half of the book profiles diverse case studies from around the world, offering context-specific insight about on-the-ground activities associated with women and wildlife trafficking.

This book will be of great interest to students and scholars of wildlife crime, environmental law, human geography, conservation, gender studies and green criminology. It will also be of interest to NGOs and policymakers working to improve efficacy of efforts targeting wildlife crime, the illegal wildlife trade and conservation more broadly.

Helen U. Agu is a Lecturer in the Department of International & Comparative Law at the University of Nigeria.

Meredith L. Gore is an Associate Professor in the Department of Geographical Sciences at the University of Maryland.

Routledge Studies in Conservation and the Environment

This series includes a wide range of inter-disciplinary approaches to conservation and the environment, integrating perspectives from both social and natural sciences. Topics include, but are not limited to, development, environmental policy and politics, ecosystem change, natural resources (including land, water, oceans and forests), security, wildlife, protected areas, tourism, human-wildlife conflict, agriculture, economics, law and climate change.

Humans and Hyenas
Monster or Misunderstood
Keith Somerville

Is CITES Protecting Wildlife?
Assessing Implementation and Compliance
Tanya Wyatt

A Political Ecology of Forest Conservation in India
Communities, Wildlife and the State
Amrita Sen

Ethics in Biodiversity Conservation
Patrik Baard

Protected Areas and Tourism in Southern Africa
Conservation Goals and Community Livelihoods
Edited by Lesego Senyana Stone, Moren Tibabo Stone, Patricia Kefilwe Mogomotsi and Goemeone E. J. Mogomotsi

Women and Wildlife Trafficking
Participants, Perpetrators and Victims
Edited by Helen U. Agu and Meredith L. Gore

For more information about this series, please visit: www.routledge.com/Routledge-Studies-in-Conservation-and-the-Environment/book-series/RSICE

Women and Wildlife Trafficking

Participants, Perpetrators and Victims

Edited by
Helen U. Agu and Meredith L. Gore

First published 2022
by Routledge
4 Park Square, Milton Park, Abingdon, Oxon OX14 4RN

and by Routledge
605 Third Avenue, New York, NY 10158

Routledge is an imprint of the Taylor & Francis Group, an informa business

© 2022 selection and editorial matter, Helen U. Agu and Meredith L. Gore; individual chapters, the contributors

The right of Helen U. Agu and Meredith L. Gore to be identified as the authors of the editorial material, and of the authors for their individual chapters, has been asserted in accordance with sections 77 and 78 of the Copyright, Designs and Patents Act 1988.

All rights reserved. No part of this book may be reprinted or reproduced or utilised in any form or by any electronic, mechanical, or other means, now known or hereafter invented, including photocopying and recording, or in any information storage or retrieval system, without permission in writing from the publishers.

Trademark notice: Product or corporate names may be trademarks or registered trademarks, and are used only for identification and explanation without intent to infringe.

British Library Cataloguing-in-Publication Data
A catalogue record for this book is available from the British Library

Library of Congress Cataloguing-in-Publication Data
Names: Agu, Helen U., editor. | Gore, Meredith L., 1977- editor.
Title: Women and wildlife trafficking : participants, perpetrators and victims / edited by Helen U. Agu and Meredith L. Gore.
Description: Abingdon, Oxon ; New York, NY : Routledge, 2022. | Series: Routledge studies in conservation and the environment | Includes bibliographical references and index.
Identifiers: LCCN 2021049258 (print) | LCCN 2021049259 (ebook) | ISBN 9780367640262 (hbk) | ISBN 9780367640286 (pbk) | ISBN 9781003121831 (ebk)
Subjects: LCSH: Wildlife crimes. | Wildlife smuggling. | Women. | Conservation of natural resources.
Classification: LCC HV6410 .W66 2022 (print) | LCC HV6410 (ebook) | DDC 364.16/2859—dc23/eng/20211202
LC record available at https://lccn.loc.gov/2021049258
LC ebook record available at https://lccn.loc.gov/2021049259

ISBN: 978-0-367-64026-2 (hbk)
ISBN: 978-0-367-64028-6 (pbk)
ISBN: 978-1-003-12183-1 (ebk)

DOI: 10.4324/9781003121831

Typeset in Bembo
by codeMantra

For our daughters

Contents

List of figures ix
List of tables xi
List of contributors xiii

1 **The Roles of Women in Wildlife Trafficking Are Mostly Unknown, But Not Unknowable** 1
   MEREDITH L. GORE AND HELEN U. AGU

2 **Using a Feminist Political Ecology Lens to Explore the Gendered Dimensions of Wildlife Trafficking Literature** 11
   CYDNEY ANDREW AND HELEN U. AGU

BOX 1
Voices from the Field: Offenders 27
HELEN U. AGU AND MEREDITH L. GORE

3 **Women as Agents of Change in Efforts to Disrupt Illegal Wildlife Trade** 30
   JESSICA GRAHAM

4 **Using a Feminist and Green Social Justice Perspective to Better Understand Governance of Wildlife Trafficking in Mexico** 44
   INÉS ARROYO-QUIROZ, JESÚS IGNACIO CASTRO SALAZAR, AND SERENA ERÉNDIRA SERRANO OSWALD

5 **Health Implications of Women's Involvement in Wildlife Trafficking in Nigeria** 59
   AMELIA NGOZI

BOX 2
Voices from the Field: Defenders   69
HELEN U. AGU AND MEREDITH L. GORE

6 Women, Wildlife Crime, and Sustainable Livelihoods in Cameroon   72
ERIC D. NANA

7 Women in the Rhino Poaching Conflict: A Feminist Political Ecology Analysis   87
FRANCIS MASSÉ, NICIA GIVÁ, AND ELIZABETH LUNSTRUM

8 Perceptions of Indigenous Baka Women's Inclusion in Wildlife Conservation and Exploitation   106
JEAN CHRISTIAN MEY BOUDOUG, HELEN U. AGU, POBO KENFACK SERGE RICARDO, AND MEREDITH L. GORE

BOX 3
Voices from the Field: Influencers   123
HELEN U. AGU AND MEREDITH L. GORE

9 Shaping Life in the Shadows: Gendered Dimensions of Wildlife Economies and Interventions in Central African Republic   126
CAROLYN A. JOST ROBINSON, ELIZABETH S. HALL, AND MELISSA J. REMIS

10 A Gendered Case File Analysis of Reptile Trafficking and Illegal Keeping in Norway   143
RAGNHILD A. SOLLUND

BOX 4
Voices from the Field: Beneficiaries and Persons Harmed   161
HELEN U. AGU AND MEREDITH L. GORE

11 Conclusion: Bridging Knowledge–Action Boundaries   165
MEREDITH L. GORE AND HELEN U. AGU

Index   169

Figures

4.1	Gender representation across environmental commissions according to Legislature, Chamber of Senators, 1997–2021	50
4.2	Gender representation across environmental commissions according to Legislature, Chamber of Deputies, 1997–2021	50
6.1	Percentages of women observed to play different roles in wildlife crime in both urban and rural areas in Cameroon in order of importance. Most women benefited directly or indirectly from wildlife crime	79
7.1	Three pillars of a feminist political ecology of poaching economies promote demarginalization of perspectives at different scales, from different lived experiences, and within multidimensional relationships	94
7.2	Much of the Great Limpopo Transfrontier Conservation Area (GLTFCA) overlaps with rhino poaching conflict at the Mozambique–South Africa borderlands	98
8.1	The Dja Biosphere Reserve in Cameroon is surrounded by several land use types, including protected areas with restricted use, hunting safari, villages, and agroforestry	110
9.1	Map of Dzanga-Sangha Protected Areas Complex, Central African Republic	130
9.2	Timeline of logging concessions in Dzanga-Sangha Protected Areas Complex. ★Indicates establishment of the protected area	132

Tables

4.1	Gender representation in the two main Deputy Attorney General Offices, PROFEPA headquarters, 2018–2020	51
4.2	Gender representation by key position in Mexico's Deputy Attorney General Offices at PROFEPA headquarters, 2018–2020	52
4.3	Gender representation by key position within Mexico's PROFEPA delegations, 2018–2020	52
6.1	Wildlife trafficking channels that show actors in different categories observed in Cameroon	74
7.1	Integrating feminist political ecology, feminist criminology, and feminist green criminology enables an intersectional approach that explores the feminist political ecology of poaching	90
8.1	Descriptive statistics of study participants (n = 300) from the Dja Biosphere Reserve	113
8.2	Mean inclusion scores for indigenous Baka women based on expert perceptions (n = 300)	114
8.3	Interview respondents' perceived involvement of women in general and indigenous Baka women (IBW) specifically in various wildlife-related activities across different micro-landscape types adjacent to the Dja Biosphere Reserve (e.g., protected area [PA], safari, and other)	115

Contributors

Helen U. Agu is a Lawyer and Senior Lecturer at the Faculty of Law, University of Nigeria. She earned her PhD in Environmental Law and Policy from the University of Nigeria, Nsukka, where she also earned her BSc, LLM, and LLB. She is an African Futures Post-Doctoral Research Scholar of the Alliance for African Partnership and Michigan State University (2019), where she conducted a multidisciplinary research on the Gendered Dimensions of Wildlife Trafficking in Sub-Saharan Africa in the Department of Wildlife & Fisheries. She was a Visiting Researcher at the Raoul Wallenberg Institute for Human Rights and Humanitarian Law, Lund, Sweden, where she researched human rights and African women's vulnerability to climate change at the Centre for Human Rights, Addis Ababa University, Ethiopia. Helen has a passion for Environmental and Conservation Law and has published papers on environmental and climate change law. She is the Director of Nature Rights Foundation – Nigerian-based conservation NGO and is currently an Education for Justice Champion for United Nations Office on Drugs and Crime, Nigeria, for her passion in piloting education for rule of law in tertiary institutions. Her current research focus is on climate change law and policy, conservation law, and gender.

Cydney Andrew recently received her MS from the Department of Fisheries and Wildlife at Michigan State University, where she also received her BS in Fisheries and Wildlife and BA in Arts and Humanities. Her research focuses on the gendered dimensions of natural resource management as well as wildlife crime. She is preparing for a professional career in policy advisory and conservation strategy.

Inés Arroyo-Quiroz is a Biologist (National Autonomous University of Mexico, UNAM, 1996), with a PhD in Biodiversity Management (University of Kent, UK, 2004) and a Postdoctoral Fellowship at the Research Institute of Ecosystems and Sustainability at UNAM, 2004–2006. She works as a Full-Time Researcher at the Regional Centre for Multidisciplinary Research at UNAM focusing on the use of animals, wildlife trade, green criminology, and cultural representations of wildlife, recently leading two international collaborations on the dynamics of wildlife trade between Mexico and the European Union. Currently, she is working on the following interdisciplinary projects: a historic perspective of the fur and leather trade from wild species involving Mexico 18th–21st centuries and representation of wild animals in Mexican cinema using green cultural criminology. Inés is active as Chair of the IUCN Green Criminology Specialist Group, a global network created to provide guidance on identifying harmful socioecological transgressions regardless of legality per se.

Jesús Ignacio Castro Salazar received a PhD in Social Sciences with a focus on Sustainable Development from the Autonomous University of Nuevo Léon and an MA in Environmental Management from the North Border College. He has worked as an inspector in the Federal Attorney General for Environmental Protection and has collaborated with the Mexican Center for Renewable Energy Research. Currently, he is a member of the National Research System and a professor–researcher at the National Technology of Mexico/ITS, Abasolo. His research focuses on environmental and natural resource management, public policy, public administration, law, and sociology.

Nícia Givá is an Assistant Professor at the Faculty of Agronomy and Forest Engineering, Eduardo Mondlane University in Mozambique. She holds a PhD in Environmental Communication from the Department of Urban and Rural Development at Swedish University for Agricultural Sciences in Uppsala-Sweden and an MS degree in Education and Training for Rural Development from Reading University in the United Kingdom. Her research scope expands from rural sociology, gender, and community-based participatory rural development to environmental politics of conservation areas, governance of protected areas, environmental communication, and sustainable natural resources management in the complex conservation–rural livelihoods nexus. She applies systemic action research, multi-stakeholders' facilitated dialogue, and social learning for collective exploration of adaptive management opportunities to reconcile conservation objectives and people's livelihood needs in inhabited protected areas in Southern Africa Transfrontier Conservation Areas.

Meredith L. Gore is an Associate Professor of Human Dimensions of Global Environmental Change in the Department of Geographical Sciences at the University of Maryland, College Park. Her research uses risk concepts to build new understanding of human–environment relationships and is designed to build scientific evidence for action. The majority of her activities can be described as convergence research on conservation issues such as wildlife trafficking, illegal logging, fishing, and mining. Meredith has conducted research in collaboration with local communities in 15 countries on 5 continents with funding from the National Science Foundation, U.S. Fish & Wildlife Service, Michigan Department of Natural Resources,

Department of Environment, Food & Rural Affairs Global Wildlife Conservation, and others. Her website hosts voluntary, consensus-based, and open-access GIS standards to combat wildlife trafficking. In 2016–2017, she served in-residence at the US Department of State Office of the Geographer and Global Issues as a National Academies of Sciences Jefferson Science Fellow. Meredith has continued to provide senior science advising to the State Department on conservation crime as an Intelligence Research Expert; she has advised the United Nations Office on Drugs and Crime's Office of Wildlife and Forest Crime, the African Union Commissions' Department of Rural Economy and Agriculture, Wildlife Conservation Society's Urban Bushmeat Team, and Michigan Department of Natural Resources' Wildlife Division. Dr. Gore has published over 75 referred journal articles and book chapters and currently serves as an Associate Editor for *Global Ecology and Conservation*. She is the Editor and Author of *Conservation Criminology* (2017). Meredith earned a PhD from Cornell University, an MA from The George Washington University, and a BA from Brandeis University.

Jessica Graham received an MS from the University of California, San Diego, in International Affairs and Environmental Policy with a China focus and a BA from the University of Florida. Serving as a Lecturer and Author on wildlife crime issues, Jessica has over a decade of experience working at the cross-section of conservation and law enforcement. She previously worked at INTERPOL and served as a Senior Advisor for the US Department of State, where she developed and led the wildlife trafficking team and championed for the US government transnational organized crime and environmental security projects. Jessica is currently the President of a woman-owned, small business consultancy, JG Global Advisory, based in Washington, DC, where she works on international environment and security topics for strategic planning and project management.

Elizabeth S. Hall is a Doctoral Candidate in Anthropology at Purdue University. Her doctoral research focuses on zoonotic emerging infectious diseases at the human–primate interface. She combines anthropological, nutritional, and epidemiological methods to investigate factors impacting novel zoonotic viral emergence. She is currently a guest researcher at the Centers for Disease Control and Prevention in Atlanta, Georgia, and serves as the technical advisor for Chengeta Wildlife in Dzanga-Sangha Protected Areas in Central African Republic.

Carolyn A. Jost Robinson is the Director of Sociocultural Research and Community Engagement for Chengeta Wildlife. She is also a Visiting Scholar at Purdue University, USA. Dr. Jost Robinson is a broadly trained anthropologist whose work emphasizes conservation in coupled human–natural systems to examine how human communities and wildlife populations respond and adapt when confronted with changing technology, human migration, and burgeoning economies of wildlife. Her research program spans the subfields of anthropology and has resulted in an innovative understandings and applications of anthropology in the biological and sustainability sciences. Dr. Jost Robinson's applied work and engagement are driven by a commitment to realize futures that include both humans and wildlife through socially responsible, locally inclusive conservation practice.

Elizabeth Lunstrum is an Associate Professor in Environmental Studies and Global Studies at Boise State University, USA. A geographer and political ecologist, her research interests include green militarization, the illicit wildlife trade, environmental displacement, and efforts to make conservation practice more inclusive of local and Indigenous communities. She has been conducting research in the Mozambican–South African borderlands since 2003.

Francis Massé is a Senior Lecturer in the Department of Geography and Environmental Sciences at Northumbria University, UK. His research takes a political–ecology and political–geography approach to understand commercial poaching and wildlife crime, the ways in which these are responded to, and the socio-economic, conservation, and ecological outcomes. He has a regional focus on Mozambique and South Africa while looking at global trends. His recent work is published in the *Annals of the American Association of Geographers, Geoforum,* and *World Development* in addition to a range of conservation journals. Francis regularly engages with conservation practitioners and policy makers to find sustainable solutions to poaching.

Jean Christian Mey Boudoug received an MS in Forest Governance (Human Dimension of Conservation), an MS in Agroforestry and Environmental Optimization, a BS in Cartography from the University of Dschang, and a Senior Forestry Technician Diploma from the National Forestry School of Mbalmayo-Cameroon. As a Senior Forestry Technician in the Department of Wildlife and Protected Areas in Cameroon, he applies forestry and GIS to the management and securing of protected areas, as well as the fight against illegal wildlife trade. Christian is President of a local organization for conservation, "The New Forestry," advocating for an inclusive management of the Mbalmayo Forest Reserve's resources. His most recent work in the Reserve is published in the *Journal of Forest and Environmental Sciences.* He plans to begin a PhD registration in Forest Governance in November 2021.

Eric D. Nana is a Senior Research Officer at the Institute of Agricultural Research for Development in Cameroon, where he leads the wildlife research unit. He is also a Visiting Lecturer at the Higher Institute of Environmental Sciences in Cameroon and a Fellow of the African Academy of Sciences. He is the President of Cameroon's Chapter of the Society for Conservation Biology, and a Post-Doctoral Fellow at the Department of Zoology of the University of Oxford. He obtained his PhD in Ecology from Charles University in Prague, Czech Republic, and his current research aims to understand the dynamics of urban wild meat markets in Central Africa. His research strives to find solutions that work best in conservation, and he is also interested in studying how ecological niches of threatened species are affected by climate change and human activities.

Amelia N. Odo is a Senior Lecturer in the Department of Human Kinetics and Health Education, University of Nigeria, Nsukka, Nigeria. She received a PhD and an MS in Public Health Education, a BS in Health Education, and Diplomas in Public Health Nursing, Midwifery, and General Nursing. Dr. Odo was a Post-Doctoral Researcher at the Department of Anthropology, Michigan State University. Her primary research interests are in sexual and reproductive health, infectious diseases prevention and control, and gender issues.

Melissa J. Remis is a Professor of Anthropology at Purdue University, USA, and the Head of the Department of Anthropology at Purdue, having served as the Associate Dean for Research and Graduate Programs in the College of Liberal Arts, 2015–2018. Dr. Remis' research in the Central African Republic originally focused on the behavioral ecology and nutrition of western gorillas, which were poorly known before she initiated her field research in the late 1980s. Her current interdisciplinary and collaborative anthropology research program combines ecological, biological, and ethnographic methods and approaches to address human–wildlife environmental relationships, conservation, gender, health, and social change in the forests of Central Africa, with publications in a broad range of flagship journals including *American Anthropologist*, *Conservation Biology*, and the *American Journal of Primatology*. She has authored or coauthored over 38 scientific articles in peer-reviewed volumes and journals such as the *Conservation Biology, American Anthropologist, American Journal of Primatology, American Journal of Physical Anthropology, International Journal of Primatology,* and *Primates*. She has been appointed to the IUCN primate specialist group and has served on the editorial board of several journals and panels of granting agencies, including the National Science Foundation. She has worked to nurture scholarly communities, improve diversity, foster inclusive research programs to expand participation among underrepresented groups, and build international field and laboratory research capacity. Remis is a recent recipient of the Violet Haas Award for contributions to the advancement of women and building a positive professional climate for women at Purdue University.

Pobo Kenfack Serge Ricardo is a Research Project Assistant at Forest and Rural Development in Cameroon. He received an MS in Agroforestry and Environmental Optimization and a Bachelor Degree in Botany from the University of Dschang, Cameroon. He applies Vegetal Biotechnology and Agroforestry Science to the multiplication and vulgarization of Bamboo-based Agroforestry Systems in southern Cameroon. Prior to this, he conducted a 14-month project with "Fondation Camerounaise Terre Vivante" on the domestication of two wild yam species (*Dioscorea* spp.) by indigenous Baka populations around the Dja Biosphere Reserve. Ricardo is also volunteer in three local organizations for conservation, and plans to begin his PhD in November 2021.

Serena Eréndira Serrano Oswald is a full-time researcher at the Regional Centre for Multidisciplinary Research, National Autonomous University of Mexico (UNAM). In addition to holding a PhD in social anthropology and doctoral studies in psychology from UNAM, she holds an MS in social psychology, an MFT in systemic family therapy, and a BA with Honors in political studies and history. She has undertaken post-doctorate studies in sociology and gender (UNAM) and postgraduate specializations in teaching and supervision, infant and adolescent psychotherapy, Gestalt psychotherapy, person-centered therapy, and gender justice and public politics, and has eight professional diplomas in social sciences, gender, and psychotherapy. She is President of the Latin American and Caribbean Regional Science Association, and was General Secretary of the Latin-American Council for Peace Research (2017–2019) and President of the Mexican Association of Regional Development (2013–2016). Her areas of interest are gender, social representations, resilience, migration, regional development, and environment.

Ragnhild A. Sollund is Professor in criminology at the University of Oslo in Norway, Department of Criminology and Sociology of Law. She has published widely within different research areas, such as policing, migration, and green criminology, particularly concerning wildlife crimes and wildlife trafficking. She has written and edited many books, including *The Crimes of Wildlife Trafficking: Issues of Justice, Legality and Morality* (2019, Routledge) and *Green Harms and Crimes, Critical Criminology in a Changing World* (2015), which received the CHOICE Outstanding Academic Title. Sollund is recipient of the Energy Globe Award and has been a partner in the EU-funded research project European Union Action to Fight Environmental Crime (2012–2016). She is currently conducting an international research project Criminal Justice, Wildlife Conservation and Animal Rights in the Anthropocene.

1 The Roles of Women in Wildlife Trafficking Are Mostly Unknown, But Not Unknowable

Meredith L. Gore and Helen U. Agu

Our Journey

We started working on this project around February 2020 when we were both in residence at Michigan State University in the United States. This was before the global coronavirus pandemic changed the world. Our science was different then; we both realized the massive hole in scientific understanding about the roles of women in wildlife trafficking around the same time that we were asked to lend our voice to a conversation and to help advance the knowledge base. What are the roles that women hold, why do they assume them, and who are the women that participate? Why do women choose to participate in wildlife trafficking or efforts to combat it? What situations surround women's inclusion or exclusion from efforts to combat wildlife trafficking? Under what conditions are women differentially impacted by the act of wildlife trafficking and programs designed to mitigate it? It was, and is clear, that scientists and others have mostly failed to consider the important causes and consequences of the gendered dimensions of wildlife trafficking. Many research questions are of interest, and some of them are addressed in this volume: What are the costs and benefits to women of intervening in wildlife trafficking along each link of the supply chain (e.g., source, transit, destination)? Is thinking about women's participation in interventions necessary or sufficient for successful programs designed to combat wildlife trafficking? Do differences between men and women signal different priorities or incentives to participate in efforts to reduce wildlife trafficking-related risks to humans and wildlife? What are the key nuances of women's participation in poaching, trafficking, and selling illegal wildlife products, either directly, in supporting roles, or as managers?

The winds for our scholarship shifted in 2020 (and most of 2021). As a team, the pandemic has been very difficult; individually, it has been devastating. Many of us cared for family members who recovered from COVID-19, and others mourned painful losses. We have shared so many emails, SMS, WhatsApp, encouraging words, and no words at all. This volume is truly a testament to the drive and passion of all the authors to use their expertise and experience for social good and to advance the conversation about women's

DOI: 10.4324/9781003121831-1

roles in wildlife trafficking. Assembling the amazing team of coauthors in this volume, and then collaborating over the course of about a year and a half, has been rewarding both professionally and personally.

Since we submitted our book proposal in early 2020, there has been growth in the topic of women and wildlife trafficking. New edited volumes by amazing scholars are forthcoming, academic activities such as webinars and conferences are being held, donors are investing, and practitioners are programming. We view this growth as positive. There is plenty of space for diverse voices.

Wildlife Trafficking and the Illegal Wildlife Trade

Wildlife trafficking and the illegal wildlife trade are not new societal challenges; however, there is broadening recognition of the problem as being an environmental crime with implications for socio-environmental security. New development programs, research grants, journal articles, programs, and policies are being implemented around the world. Similarly, wildlife trafficking and the illegal wildlife trade are criminalized increasingly. For example, Thailand's 2019 Wildlife Conservation and Protection Act B.E. 2562 imposed minimum fines, increased maximum fines 25-fold from previously published guidelines, and increased maximum imprisonment terms 2.5–3.5-fold, depending on species (USAID, 2019). Uganda's Wildlife Act of 2019 legislated 10 years imprisonment or 350 currency points or both for a first offense of a general penalty under the Act and 20 years or 500 currency points or both for a second offense. Offenses against protected species were increased to life imprisonment or 10,000–1,000,000 currency points or both (Uganda Wildlife Act, 2019).

Increased criminalization of wildlife trafficking and illegal wildlife trade has paralleled an increase in attention from criminologists, computer scientists, operations engineers, and environmental justice scholars. There is a growing, dynamic literature on the topic; a broad discussion is beyond the focus of this book, but the case studies in this volume help illustrate the scope, scale, and impact of the problem on the ground and in different contexts. The conflict theory of crime stresses that definitions of crime and the application of the label 'criminal' are determined by the interests of those with economic and political power (e.g., Eliason, 2020). Power, economics, culture, and justice are all topics that permeate discourse and science about wildlife trafficking and the illegal wildlife trade. But we should and can know more about the conditions under which wildlife trafficking involves violence against people, animals, and the environment. How does wildlife trafficking undermine the rule of law, degrade culture and ecosystems, sustainable development investments, and legal supply chains, link to spread of zoonotic disease and gender-based violence, and magnify existing structural injustices or create new ones (Brito et al., 2018; Kahler & Gore, 2012; Gore et al., 2019; Gruber, 2017; Smith et al. 2017)?

The science of wildlife trafficking is accentuated by the borderless nature of the crime, diversity of wildlife impacts, and its effect on human communities. Perhaps, in part, because wildlife trafficking can involve multiple serious aspects of criminality, violence, and violations of the rule of law, it is viewed (1) as a crime and (2) worthy of interdisciplinary and multi-sectoral investment, even by the scientific community. Society's efforts, to date, have failed to consider the important causes and consequences of the gendered dimensions of wildlife trafficking. Thus, policies, programs, and projects designed to reduce risks that are associated with wildlife trafficking that involve women may fail to achieve outcomes, fail to be monitored and evaluated using accurate metrics, and fail to promote the voice of local scientists during science–policy discourse.

Beginning to Explore the Unknown

This is the first volume of which we are aware that integrates diverse perspectives, studies, and analyses on the roles of women in wildlife trafficking. Thus, there is *a lot* of diversity contained within these pages. Below are three considerations to remember as one begins to "explore the unknown." These considerations permeate all chapters in this volume and offer multi-level perspectives about the roles of women in wildlife trafficking: individual women, groups of women, or women in conservation or law enforcement organizations. We believe that wildlife trafficking constitutes a globally distributed problem with broad socio-environmental significance. Some readers may find these considerations to be tautological or obvious, but we believe they form a foundation for inquiry and serve as a helpful reminder of context. This is because the roles of women in wildlife trafficking have been mostly neglected by scientists (and others). Although the neglect of women by criminologists has been well documented, for example, some suggest it reflects the hierarchical structure both in society and in the discipline that places men in positions of power and diminishes the importance of women (e.g., Moyer, 1992).

1   *Women assume roles in both their private and public lives.* Around the world, women (and men) inherently play different roles in their public and private relationships, and our knowledge about each is not equal. There may not always be a difference in these roles, but the different places and spaces are worthwhile to consider because they can influence how, when, and why we derive the knowledge that we do about those roles. Scarborough et al. (2002), for example, studied the roles of women in public and private law enforcement organizations. Their book reminds us to consider this duality explicitly, or perhaps even a spectrum of roles, when thinking about the topic. *Representation* of women is just one way that the two spaces may differ. Recruitment and retention issues into different roles remain mostly unknown, although women can be represented

across different role types such as offender, defender, influencer, victim, beneficiary, and observer (e.g., see Agu & Gore 2020). Scarborough et al. (2002) noted that public occupations are better studied than private occupations or roles; research exists, for example, on biased entry tests, discrimination, sexual harassment, lack of mentoring programs, targeted recruitment of men, and outdated models of police work.

Both women and men are assigned gender role expectations, particularly what is perceived as being appropriate behavior, by social norms that vary geographically. For example, law enforcement is a traditionally masculine occupation, and occupational subcultures may create barriers to women being accepted into law enforcement positions. Further, men and women have different expectations about what constitutes appropriate behavior in public compared with private spaces (e.g., choice of clothing, choice of livelihood), regardless of whether or not these spaces are for employment. Gender role expectations are especially relevant when women pursue (and succeed and fail) non-traditional occupations or roles. Studies on this topic traditionally focus on sex-role expectations. Historically, it has been suggested that women reflect the stereotypical behavioral feminine characteristics of being weak, passive, and nurturing. As women gain entrance into non-traditional occupations or roles, sex-role expectations may affect opportunities and retention. To succeed, they must justify their existence in the organization and have strong reasons for their choice of career or role.

Scarborough et al. (2002) reviewed principles of occupational socialization, which overlay and interact with other social norms and gender role expectations. Occupations such as law enforcement and conservation have their own subcultures that have their own distinct characteristics. Law enforcement subcultures have been characterized as being very distinct due to the nature of the work that sets them apart from most other occupations. The more one strays from a gender-appropriate occupation, the more severe the consequences can be for that individual's social, cultural, and even physical well-being.

2 *Women can be viewed as outsiders and, therefore, be minoritized.* The fields of criminology and criminal justice have helped to delineate the extent to which men and women differ in their occupational/role strategies, what the occupational/role costs are, and when differences are delineated or change. Often, research on women in law enforcement, for example, is focused on issues of competency, attitudes toward other women in policing, work-related stress, legal issues, or descriptions of current populations of women. Exploring "insider–outsider" relationships has implications for understanding the variety of roles that women play in the law enforcement sector (e.g., as defenders or police) or in wildlife trafficking (e.g., as offenders). Where do they succeed and where they struggle (see Lute & Gore, 2014 for a non-gendered discussion of ingroup–outgroup relationships in wildlife conservation)? These insights

are important because we know some women may get entrapped in a role and experience stereotyping, isolation, performance pressures, frustration, and dissatisfaction that is frequently experienced in policing. We do not know how women cope with being outsiders and the 'us' versus 'them' dichotomy in wildlife trafficking roles. Wexler (1985) identified five different "role styles" that female police officers used in their roles: neutral, impersonal, semi-masculine, feminine, and mixed. We do not know how role styles apply in the wildlife trafficking space or what the implications are for those expressions of role styles.

People who are "new" to a particular role, who may be women in a traditionally male-dominated occupation, often challenge the solidarity of the group and create status conflicts (i.e., intergroup conflict) (Lute & Gore, 2014). Oftentimes, a way of dealing with a status challenge is to shun contact with new people or to put them in places in which they are not visible to the public, for example. Women are often marginalized and minoritized by their professional or socio-cultural/ethnic role status, but their status also gives them a unique perspective with which to see an organization. Because of their outsider-within status, some women have a distinct perspective that is different from the dominant group's perspective. Subordinate groups can use their outsider-within status creatively as a source of ideas. Women in non-traditional occupations exemplify the outsider-within perspective. We do not know how this outsider-within personification influences the policing of wildlife trafficking.

3 *Women are treated as unique beings, some of the time*. Women can be unique because context-specific social changes and demands of interest groups have created "pressures" for specialized handling of certain types of victims (e.g., of sex offenses). With these "pressures" comes the notion that women in particular roles or occupations, such as law enforcement officers, have special skills for dealing with victims. This thesis challenges, for example, the male officer's capabilities of handling such cases and creates roles for female police officers to solve dilemmas for men. Indeed, changes in the role of women in law enforcement and conservation often require legislation or policy changes, executive/leadership action, judicial decisions, and altering eligibility criteria and selection standards (Moyer, 1992). This women-as-unique approach has a long tradition, where women have long been seen as more virtuous until they make a mistake, intentionally or unintentionally. Females who deviate from their expected roles and transgress the law, for example, may be viewed as immoral, corrupt, hysterical, diseased, manipulative, or devious.

Moyer (1992) delineated three models that highlighted how being a woman impacted the processing of female offenders by law enforcement authorities: (1) being a woman was not a factor in the treatment of women, (2) women received preferential or chivalrous treatment, or (3) women were treated punitively. Among those three models, the chivalrous model has received a lot of scientific attention largely because women have low crime rates. Thus, women may be viewed as being less capable than men of committing criminal acts. Or sex differences in crime seem related to the different roles of men and women and the different social worlds in which they live. Moyer (1992) noted that sometimes there was actual participation of women in crimes, but not of a type that led to an arrest.

Definitions and Conventions

This volume explores woman *and* wildlife trafficking *and* the illegal wildlife trade, which we conceive of as being identical twins – distinct individuals born from the same parents and looking almost the same. Not all wildlife trade is illegal and not all illegal wildlife trade crosses geopolitical boundaries to become wildlife trafficking. Wildlife trafficking and illegal wildlife trade are not the only types of wildlife crime. The differences are important, but beyond the scope of this volume. We invited authors to use their own definitions based on their own lived experiences and expertise, but we asked them to be consistent in their chapters.

This volume also explores women. We apply definitions from the Johns Hopkins University Gender Analysis Toolkit (Jhpiego, 2020). Thus:

> **Sex** refers to biologically defined and genetically acquired differences between males and females, according to their physiology and reproductive capabilities or potentialities. It is universal and mostly unchanging, without surgery. **Gender** refers to the economic, social, political, and cultural attributes and opportunities associated with being women and men. The social definitions of what it means to be a woman, or a man vary among cultures and change over time. Gender is a sociocultural expression of characteristics and roles that are associated with certain groups of people with reference to their sex and sexuality. **Gender Stereotypes** are ideas that people have on masculinity and femininity: what men and women of all generations should be like and are capable of doing.

Empowerment and Mainstreaming

Advancing scientific insight about the role of gender equality, and thus women, in wildlife trafficking is crucial for sustainable conservation (Sharma, 2017). The contributions of women to the varied dimensions of wildlife trafficking (and other natural resource contexts) are mostly unrepresented in official statistics and thus are invisible. Women are mostly recognized as helping or supporting men's wildlife trafficking-related activities, even when women participate or lead. The low number of recognized female roles and contributions may be one factor driving exclusion of women from wildlife trafficking-related policies and programs. Research in other sectors (e.g., sustainable fisheries management) has confirmed such an influence of exclusion (Torre et al., 2019). It is not only the focus on women and equality that is missing from wildlife trafficking-related data sets and discussions, but a stronger focus on knowledge, vision, fairness, governance, gender balance, and creative solutions to addressing the problem of wildlife trafficking. A lack of evidence about gender may reinforce the notion that women are not involved, which further excludes women from decision-making forums or processes (Sharma, 2017). This volume aims to help fill this gap with discussions on theory and case study illustrations, with a key implication being enhanced empowerment.

Empowerment is defined by some as the capacity to make strategic life choices (Kabeer, 1999), particularly in contexts where choices were previously denied. Three commonly considered dimensions in women's empowerment are resources (pre-conditions), agency (process), and achievements (outcomes). With regard to resources, women may receive skills training and equipment to participate in wildlife trafficking-related and other conservation activities; they may also receive training in leadership, negotiation, and decision-making. Key to this resource dimension is equal pay to men, as well as equal access, permits, and external financial support. With regard to agency, women may start their own initiatives, seek and obtain investments, hire employees, and engage in specific partnerships. Finally, with regards to outcomes, knowledge acquired from women through new roles can provide opportunity to participate in decision-making on the same level as men. In this regard, women can influence decisions about their future (Sharma, 2017). Having female role models and men that actively support professional development of women further advances women's empowerment. Empowered individuals and communities may more effectively resolve socio-environmental problems associated with wildlife trafficking.

Empowerment may also be defined as an "inclusive process of encouraging, enabling, and developing the capability for self-sufficiency, self-dependence, self-assertion and autonomy of marginalized and disempowered groups or community through consciousness-raising, proactive participation in public life and mobilization for right entitlements" (Sharma, 2017). This conceptualization is thought-provoking for wildlife trafficking because it raises questions about disempowerment, both in private (e.g., family, peer group) and public spaces (e.g., school, club, community). Sharma (2017) argued notions of empowerment and disempowerment are fundamentally about male–female relationships; these relationships have multiple dimensions: biological dimensions take the form of difference in some parts of the body; spatial or geographical dimensions take the form of distancing; social cultural dimensions take the form of discrimination; economic dimensions take the form of deprivation; and political dimension takes the form of disempowerment. Disaggregating these dimensions can result in practical insights for structuring, monitoring, and evaluating efforts to mainstream gender into decisions about wildlife trafficking, for example, across source, transit, and destination geographies.

Mainstreaming offers a strategy for making women's and men's concerns and experiences a central dimension of the design, implementation, monitoring, and evaluation of policies and programs so that women and men benefit equally, and inequality is not perpetuated (United Nations Economic and Social Council, 1997). Gender mainstreaming actions often include gender training, gender analysis, or women's empowerment activities (Ogra, 2012). Although there is evidence that gender mainstreaming results in improved conservation and development outcomes in general, the practice is inconsistently implemented (Ogra, 2012). There is virtually no evidence in wildlife trafficking contexts, precluding evidence-based decision-making. Scientific evidence could help advance the scope and scale of gender mainstreaming efforts across the landscapes within which wildlife trafficking operates.

Organization of the Book

This volume is organized into two sections. The first half assesses the state of science, offering four distinct perspectives on how women and wildlife trafficking can be explored by students of the topic. Not only are distinct conceptual frameworks offered in different chapters, but the varying disciplinary vistas enabled by distinct frameworks are explored. In Chapter 2, the strengths and weaknesses of feminist political ecology are explored within the context of women and wildlife trafficking. The chapter reviews the literature on wildlife trafficking that discusses, demographically and theoretically, who participates. Chapter 3 conceptualizes women as agents of change, or individuals that may initiate, motivate, or manage shifts in the wildlife trafficking status quo. The chapter reviews four different roles for women and profiles how each has seized opportunity previously to engender change. Chapter 4 uses a feminist historical analysis to track wildlife trafficking as a green social justice issue in Mexico, which considers the political and institutional sphere of representation in socio-cultural arenas of recognition. Chapter 5 considers the health implications of women's involvement in wildlife trafficking. Three distinct, yet related, dimensions of health – physical, social, and mental – have implications for the diversity of roles that women may play in wildlife trafficking (e.g., offenders, defenders, influencers, observers, victims, beneficiaries).

The second half of the book profiles various case studies from around the world, offering context-specific insight about on-the-ground activities associated with wildlife trafficking. These chapters apply many of the conceptual frameworks discussed in the first half of the book. Chapter 6 explores women's roles across three channels of wildlife crime in Cameroon, comparing and contrasting results from market chain and cost-benefit analyses to describe trafficking routes, control points, and sale points. Chapter 7 offers an intersectional approach to study feminist political economies and militarized responses to rhino (*Diceros bicornis*) poaching in Mozambique. The analysis scales gendered impacts from the household to the community and explores women as change agents in rhino poaching economies. Chapter 8 explores notions of indigenous women's inclusion in wildlife conservation and exploitation in Cameroon. An innovative analysis integrates landscape and human geography to determine degrees of indigenous women's inclusion across different micro-landscapes. Chapter 9 uses a historical-ethnographic approach to consider shifting wildlife economies in Central African Republic, which profiles how women have shaped their daily livelihoods in response to shifting informal economies that involve wildlife. Finally, but certainly not least, Chapter 10 applies hegemonic masculinity and case file analysis to reptile keeping in Norway, which tracks the activity over time as regulatory and legal changes influenced offenders' motivations and perceptions of animal harm.

Interspersed throughout the authored chapters are four "voices from the field" sections. These excerpts discuss different roles for women in wildlife trafficking. Direct quotations from male and female experts provide first-person perspectives on women's involvement in various wildlife trafficking-related roles, including offenders, defenders, influencers, persons harmed, and beneficiaries. The opinions offer significant and direct source material complementing the analyses and evaluations of chapter authors. They are provided as a tool to help illustrate the heterogeneity of the topic. Excitingly, the direct quotations provoke new questions that can be asked and answered by our community.

Our book is edited with both a compassionate eye for the reader (and a very patient publisher!) and for an authentic presentation of authors' expertise and lived experiences. We asked ourselves, what use is an edited volume that distills away the epistemological and ontological orientation of the authors? The content curated in this book is adventurous in its voice because all the authors are different. Although the diversity of the content may be challenging to read in total, we believe that in the end and taken together, diversity is the volume's greatest asset. Just because something is not known does not mean that it is unknowable. We hope that when readers finish the volume, they recognize how much more there is to learn about women and wildlife trafficking AND how many forms of science there are for exploring those unknowns.

This volume would not be possible without the support of our daughters. We also acknowledge Cydney Andrew for providing excellent and consistent support throughout our editing process. The Michigan State University Alliance for African Partnership supported Dr. Agu's initial fellowship, and Jamie Lake helped us navigate logistics from afar. Experts in The United Nations Office on Drugs and Crime Global Programme on Forest and Wildlife Crime helped us develop our ideas. Finally, this volume would not be possible without the significant contributions of the reviewers. Thank you so much for your professional, financial, and emotional support!

Stronger together,
MLG and HUA

Literature Cited

Agu, H. U., & Gore, M. L. (2020). Women in wildlife trafficking in Africa: A synthesis of literature. *Global Ecology and Conservation, 23,* e01166.

Brito, J. C., Durant, S. M., Pettorelli, N., Newby, J., Canney, S., Algadafi, W.,... & Carvalho, S. B. (2018). Armed conflicts and wildlife decline: Challenges and recommendations for effective conservation policy in the Sahara-Sahel. *Conservation Letters, 11*(5), e12446.

Eliason, S. L. (2020). Poaching, social conflict, and the public trust: Some critical observations on wildlife crime. *Capitalism Nature Socialism, 31*(2), 110–126.

Gore, M. L., Braszak, P., Brown, J., Cassey, P., Duffy, R., Fisher, J.,... & White, R. (2019). Transnational environmental crime threatens sustainable development. *Nature Sustainability, 2*(9), 784–786.

Gruber, K. (2017). Predicting zoonoses. *Nature Ecology & Evolution, 1*(4), 1–4.

Jhpiego. (2020). Seven steps to a gender analysis. https://gender.jhpiego.org/analysistoolkit/seven-steps-to-a-gender-analysis/.

Kabeer, N. 1999. Resources, agency, achievements: reflections on the measurement of women's empowerment. *Development and Change,* 30: 435–464.

Kahler, J. S., & Gore, M. L. (2012). Beyond the cooking pot and pocketbook: Factors influencing noncompliance with wildlife poaching rules. *International Journal of Comparative and Applied Criminal Justice, 36*(2), 103–120.

Lute, M. L., & Gore, M. L. (2014). Stewardship as a path to cooperation? Exploring the role of identity in intergroup conflict among Michigan wolf stakeholders. *Human Dimensions of Wildlife, 19*(3), 267–279.

Moyer, I. L. (1992). *Changing Roles of Women in the Criminal Justice System: Offenders, Victims, and Professionals.* Prospect Heights, IL: Waveland Press.

Ogra, M. V. (2012). Gender mainstreaming in community-oriented wildlife conservation: Experiences from nongovernmental conservation organizations in India. *Society & Natural Resources, 25*(12), 1258–1276.

Scarborough, K. E., Collins, P. A., & Scarborough, K. (2002). *Women in Public and Private Law Enforcement.* Butterworth-Heinemann.

Sharma, S. (2017). The dynamics of women's empowerment: A critical appraisal. *Social Change, 47*(3), 387–405.

Smith, K. M., Zambrana-Torrelio, C., White, A., Asmussen, M., Machalaba, C., Kennedy, S.,... & Karesh, W. B. (2017). Summarizing US wildlife trade with an eye toward assessing the risk of infectious disease introduction. *EcoHealth, 14*(1), 29–39.

Torre, J., Hernandez-Velasco, A., Rivera-Melo, F. F., Lopez, J., & Espinosa-Romero, M. J. (2019). Women's empowerment, collective actions, and sustainable fisheries: Lessons from Mexico. *Maritime Studies, 18*(3), 373–384.

Uganda Wildlife Act. (2019). https://www.informea.org/sites/default/files/legislation/Wildlife%20Act%2C%202019%20-Gazetted%20Version.pdf.

United Nations Economic and Social Council. (1997). Gender Mainstreaming. Economic and Social Council Report A=52=3. http://www.un.org.womenwatch/daw/csw/GMS.PDF.

USAID Wildlife Asia. (2019). USAID facilitates groundbreaking changes to Thailand's national wildlife law. https://www.usaidwildlifeasia.org/news/all-news/usaid-facilitates-groundbreaking-changes-to-thailand2019s-national-wildlife-law.

Wexler, J. G. (1985). Role styles of women police officers. *Sex Roles, 12*(7–8), 749–755.

2 Using a Feminist Political Ecology Lens to Explore the Gendered Dimensions of Wildlife Trafficking Literature

Cydney Andrew and Helen U. Agu

Wildlife crime is the act of violating a national or international law for the harvest, possession, or trade of wild flora or fauna. Wildlife trafficking, a type of wildlife crime, is the illegal movement and trade of wildlife, their parts, or products made from wildlife. Wildlife trafficking is sometimes estimated as the fifth largest illicit trafficking industry, which was estimated to produce $23 billion in annual revenue globally, which did not include the illegal fishing industry that produced $50 billion annually or illegal logging that produced $100 billion annually (UNODC, 2016). The wildlife trafficking industry is sometimes intertwined with other crime industries; it is known to be trafficked alongside drugs and weapons, violates labor conditions, and can be detrimental to human health when wildlife trafficking is associated with zoonotic disease transmission or transactional sex that results in sexually transmitted diseases (Basu, 2013; Fiorella et al., 2015; Mossoun et al., 2015; UNODC, 2016).

Considering these facts, nations, international organizations, foundations, and non-governmental organizations have spent millions of dollars in aid and have incorporated wildlife trafficking plans into economic, development, and conservation policies (e.g., Gore et al., 2019; World Bank Group, 2016). An increasing amount of research has been conducted on wildlife trafficking, which include what species are being trafficked, where species are being transported, and who is driving demand for trafficked wildlife. However, the gendered dimensions of wildlife trafficking are often overlooked as an important factor when it comes to explaining "who" participates in wildlife trafficking; thus, many other answers to questions about "how" and "why" and "where" wildlife trafficking occurs are also overlooked. The underdeveloped knowledge base suggests that there is gender blindness, negligence, or apathy.

A normative view on gender has helped to perpetuate the essentialist myth that women harbor an innate connection with nature and are more prone to perform conservation-conscious acts or the idea that women are less capable of committing crime and corruption and are victims as opposed to perpetrators of crime. Due to the sociocultural reinforcement of this normative view, the experiences of nearly half the global population are ignored, which

DOI: 10.4324/9781003121831-2

exclude key perspectives on wildlife trafficking that are vital to sustainable solutions. Although the dominant public perception may be that women are less likely to participate in wildlife trafficking due to their connections with the environment and perceived caregiving nature, it is impossible to ignore the fact that science has been a historically male-dominated endeavor. Today, it is reasonable (prescient!) to consider that women may experience crime differently from men – socially, culturally, and economically – and that each gender experiences different barriers and access points to any criminal industry.

The focus of this chapter is to review the current literature on wildlife trafficking about who participates in illicit wildlife trafficking. Drawing on insights from feminist political ecology and green criminology, extant literature can be analyzed with a lens on gendered dimensions of wildlife trafficking, particularly how the literature addresses the experiences of men and women vis-à-vis entry and participation into this illicit economy. Focusing on how women's experiences are presented and characterized in the current literature can help uncover new research questions. Gaps in knowledge represent outdated, underexplored, or under considered science that, in turn, can help to inform decision-making by potential donors about investing resources, policy evaluation, or program development.

Applying Feminist Political Ecology, Green Criminology, and Demographics to Wildlife Trafficking

Feminist political ecology is a conceptual framework that examines the nexus among gender, politics, and ecology that theorizes that gender is a key variable in how a person interacts with and understands the natural environment in which they live. The theoretical framework expands the scope of analysis for ecopolitical social research to include the household before the 'local' or 'community' level (Rocheleau, 2008). Within feminist political ecology, the intersectionality between gender and class, race, and culture is accentuated throughout the process of inquiry, particularly when considering access to ecologically sustainable livelihoods and the abilities that actors in a socioecological system have to respond to changes and developments (Rocheleau, 1996). Exploring these sociodemographic dimensions aids in revealing the different motivations and specific roles that individuals of different backgrounds have in participating in wildlife crime (i.e., a wealthy woman of the majority race likely traffics wildlife for a much different reason than a poor, minority man). Revealing such roles and motivations can assist defenders (as defined in Agu & Gore, 2020) in developing solutions that are directly combative to the offender. A general, one-size-fits-all approach to combat wildlife crime is not effective.

The goal of feminist political ecology is to apply gender to all parts of the research process from design to implementation and to consider gender as not only a factor of the solution, but sometimes, part of the problem as well (Rocheleau, 1995). Applying gender to all parts of the research project is not as simple as including sex and gender data as a primary form of analysis (as discussed later in this chapter), but also utilizing feminist methods, considering who comprises the population within the study area chosen for the research, and even ensuring results are readily available and are understood by all individuals who are affected by the research. When conducting studies on wildlife trafficking, this may look like choosing a gender-neutral study area (as opposed to a mining or logging camp, which are culturally male-dominated, e.g., Spira et al., 2019). Feminist methods that can be utilized are often qualitative, such as interviews and focus groups as popular choices, because qualitative analysis allows the researcher to consider the social, gendered knowledge that cannot be proven by the scientific method alone (Leung et al., 2019).

History of Feminist Political Ecology

The ideals of feminist political ecology have been laid out by a multitude of feminist and ecological scholars, among the most famous being ecofeminists Plumwood (1993), who wrote a feminist critique on the Western male domination of nature and articulated the connections among gender, race, class, and nature. Shiva (1988) drew connections between ecological destruction, colonialism, and the oppression of women based on her lived experiences in India. Merchant (1980) wrote that the Scientific Revolution (16th–17th centuries), and consequently the Age of Enlightenment (17th–18th centuries), gave rise to an ecologically destructive socioeconomic order that subordinated women. Feminist political ecology truly emerged in the mid-1990s when Rocheleau published her paper "Maps, Numbers, Text, and Context: Mixing Methods in Feminist Political Ecology" (1995), which was a case study in the Dominican Republic on feminist geography. She then published her book *Feminist Political Ecology: Global Issues and Local Experiences* (1996), which explored gendered experiences in a variety of economic and ecological communities across the globe. However, Rocheleau herself credited Blaikie and his book *The Political Economy of Soil Erosion* (1985) as being the root of her ideology and her feminist political ecology in general (Rocheleau, 2008). Beyond these essential works, other prolific scholars, such as Nightingale, advanced exploration of the ways gendered bodies used natural resources differently, primarily in community forestry initiatives in Nepal where the use of managed lands depended on the caste or gender of an individual (Nightingale 2003, 2006, 2010; Nightingale & Sharma, 2014, Nightingale et al., 2018).

Feminist Political Ecology, Crime, and Corruption

Feminist political ecology has been applied to criminological issues, particularly corruption and organized crime. Goetz (2007) explored the relationship between women and corruption and stated that women experienced corruption differently than men because one's gender affected entry into systems of corruption, ability to offend, and protection from victimization of corruption. Goetz helped to disprove the universal truism that women were less likely to participate in corrupt activities than men. Aside from Howson (2012), who applied this thinking to evaluate Senegalese women's roles in crimes related to natural resources, there has been little work on the connection between feminist political ecology and environmental crime in general. Even more sparse are the connections between feminist political ecology and wildlife trafficking. Hübschle is well known for her work in informing environmental governance against organized crime and in increasing security and restorative justice initiatives in African contexts. She dissected women's roles in all forms of crime in South Africa and noted that engaging in wildlife trafficking was prolific in some instances (Hübschle, 2014). She discussed the importance of considering the intersectionality of gender, class, and culture by analyzing wildlife-related crime because it helped to produce a more holistic understanding of organized crime networks.

Origins of Green Criminology

Historically, mainstream criminology has ignored harms against non-humans and environmental crime, which in general is a "niche interest" among criminologists (Wellsmith, 2011, p. 126). In 1990, Lynch introduced green criminology, which blended environmentalism, radicalism, and humanism to construct a novel perspective for criminological studies. Green criminology was framed as being responsive to rising environmental movements and was useful for tackling contemporary issues with a nexus to crime, such as climate change, pollution, corporate malfeasance, and animal welfare. Green criminology advanced principles from environmental sociology and critical criminology to spotlight environmental issues, such as ecosystem pollution and destruction (Eman et al., 2013). South (1998) further advanced green criminology by reviewing past studies that made contributions to the developing field and introduced ten areas where environmental/green issues intersected with criminology. These ten areas advanced the field and its application and included: (1) rethinking definitions of what constitutes harm and crime to include a green perspective; (2) enhancing capacity to enforce compliance with environmental regulations; (3) taking "old crimes and new violations" (p. 225), such as dog fighting, seriously; (4) opening the possibility for interdisciplinary work between the social and hard sciences; (5) studying organized crime and its relationship to corporate entities and poor environmental practices; (6) identifying further areas for studying 'gray policing,' defined as the cooperation between social control agencies (Hoogenboom, 1991); (7) connecting green crimes with other forms of global crime, such as crimes of war; (8) connecting green criminology with social movements, such as the construction of identities; (9) examining the future potential of damage caused by environmental crime, otherwise described as 'futurology'; and (10) bringing new ideas and directions to criminology (South, 1998).

Intersectionality of Green Criminology and the Problem of Wildlife Trafficking

Green criminology does not advocate for a definitive causal relationship between, or a specific solution to, environmental harms through crime. Instead, the framework is an argument that the field of criminology should consider the extent and implications of the harm caused to the global populace through crime (South et al., 2014). As South stated in his seminal work, "a criminology relevant to the [20th] century should… be able to embrace environmental, human, and animal rights issues as related projects. To 'think through' the implications of a green perspective, we may need some new theoretical tools and practical propositions" (1998, p. 225). Many writers in green criminology have focused their time on exposing specific environmental harms and have consequently provided detailed analyses on the processes and effects of such harms, such as pollution and the illegal trade of animals. Green criminologists attempt to explain environmental harm at both the micro- and macro-scales by utilizing prominent criminological theories, such as deterrence theory, situational crime prevention, and strain theory, to strengthen both green criminology and subfields of traditional criminology (Brisman, 2014).

Green criminology grew in scope and application, and it included an analysis of the interdependence and relationship between what was termed human and nonhuman animal interests. It offered novel opportunities to explore wildlife crime, such as bringing environmental justice perspectives into policy evaluations, harms against animals from biodiversity loss, and evaluating regulatory processes designed to protect the environment from future harms. Contemporary green criminological research has framed how to measure and to characterize wildlife trafficking, for example, by describing the attributes of illegal wildlife markets and why such markets flourish (Cao Ngoc & Wyatt, 2013). Complimentary scholarship has delineated three categories of harm caused by wildlife trafficking: harm to the environment through loss of biodiversity and transmission of zoonotic diseases, harm to animals through abuse during harvest and smuggling, and harm to the security of nations and their people through corruption, terrorism, and organized crime (Wyatt, 2013).

Women's Roles in Wildlife Trafficking Using Demographics

Demographic analyses often measure variables such as age, class, race, ethnicity, gender, and sex to gain a basic understanding of a population's composition and how it may affect the phenomena that is being studied, such as crime, educational attainment, or health security (US Census Bureau, 2016).

Feminist political ecologists have used demography to disaggregate data to understand the legal landscape of environmental actors invested in an issue, to analyze the socioenvironmental interactions between actors and regimes, and to uncover the exclusion of marginalized populations (Gillespie & Perry, 2019). Disaggregating populations can help to compare men and women to reveal where there are gaps and inequalities that are likely to affect women's participation rates, leadership, access to services, or that subject men and women to differential risks and vulnerabilities that affect their health. Gendered demographic analysis can provide an understanding about why gaps and disparities exist and how they affect women's aspirations and opportunities. Policy-relevant questions can be asked and answered, such as how the identified gender differences limited or facilitated desired changes in knowledge, practices, and access to resources in wildlife trafficking. Or what are the gender-based constraints and opportunities that have the potential to either impede or facilitate achievements in anti-poaching activities?

Although disaggregating demography can enhance understanding about the human dimensions of wildlife trafficking, the wildlife trafficking literature mostly lacks explicit demographic analyses. For example, Mbete et al. (2011) conducted a thorough analysis of household-level wild meat consumption in Brazzaville, Republic of the Congo. Oftentimes, wild meat was sourced from the illegal wildlife trade because species such as monkeys and pangolins were protected species under law and could not be served as wild meat. Mbete et al. (2011) focused primarily on ethnicity as a primary factor in an individual's participation in wild meat consumption and was not discussing the legality of wild meat explicitly. The authors surveyed participants and measured demographic variables that included ethnicity, income, household size, employment, and religion. The study did not consider gender, and there was no mention of whether the sex of participants was measured. This gap was unfortunate because analysis about the head of household's gender could have provided more detailed understanding about the relationship between ethnicity and wild meat consumption and household-level decision-making about wild meat consumption.

Research that does offer a disaggregation of study participants' gender could further advance the gendered dimensions of analyses for the benefit of gender inclusive conservation. For example, Ordaz-Németh et al. (2017) studied the effects of the 2014–2016 West African Ebola outbreak on wild meat consumption in Liberia. The study measured many demographic variables, such as the household education level, number of occupants in the home, income, perceptions of law enforcement, and property distance to roads and settlements. The study also measured participant sex, where they hypothesized that women were consuming a smaller quantity of meat and fish products, but more roots and tubers, than men. Sex was statistically insignificant in the models tested, and the authors elaborated on their findings and their implications for wildlife conservation or public health. It is not reasonable to assume that women and men are always different in their perceptions of wildlife conservation issues, which include wildlife trafficking. Empirical evaluation and rigorous interpretation of results is possible, and yet it is underexplored. Gore et al. (2016), Pires et al. (2016), Reuter et al. (2016), Tagg et al. (2018), and Spira et al. (2019) all exemplify studies that collected demographic information on sex and gender, but failed to analyze or to interpret results through a gendered lens.

Although gender is not a widely studied demographic variable in wildlife conservation, ethnicity is. For example, Poulsen et al. (2009) delineated wild meat consumption by household and ethnic group; Fa et al. (2016) compared the impact of wild meat harvest by Pygmy with non-Pygmy hunters; Wright and Priston (2010) analyzed wild meat hunting practices of different ethnic groups in Cameroon; and Cronin et al. (2017) explored the biological, socioeconomic, and cultural facets of bush meat hunting in Bioko Island, Equatorial Guinea. Each of these studies included demographic analysis, but did not discuss the sex or gender of research participants. They also did not state explicitly that (nearly) all respondents were men, and they made policy recommendations based on cultural and ethnic results. Feminist political ecology might critique these studies in their failure to consider women's perspectives. Women comprise half the population and, therefore, male-only studies, by design, overstate male perspectives. Excluding gender from demographic analysis also undermines the interpretation of results. Ethnicity and culture are influenced by and, in turn, influence gender in general and could influence results about engagement in or support of wildlife conservation (in this case, illegal wild meat trade).

Epanda et al. (2019) interviewed heads of households in the Dja Biosphere Reserve, Cameroon, to explore local perceptions of livelihood alternatives that were designed to reduce wildlife poaching. The researchers explicitly stratified their sample and interviewed households led by both men and women due to their known differentiation in livelihood strategies. The researchers reflected on the importance of gender disaggregated data in both the results and discussion sections of their article, and they noted the importance of women's membership and active participation in non-timber forest product (NTFP) industries, such as cocoa farming. Lowassa et al. (2012) studied women's influence on male participation in wild meat hunting in communities in the Lower Omo, Ethiopia, and Western Serengeti, Tanzania. Gender-disaggregated focus groups in each study area produced rich qualitative data about the nature of influence through both verbal and non-verbal communication, which included singing songs of praise, rewarding successful hunters with gifts or special foods, and romance. Men who chose not to hunt may have experienced wives who cheated, and they were called a "woman" as an insult (or being called other insulting names). Gore and Kahler (2012) interviewed members of community conservancies in Zambezi, Namibia, to gain insight on gendered perceptions of risk that were associated with human–wildlife conflicts. Gender parity was maintained during data collection and gendered differences in risk perceptions were quantified, particularly those associated with crop damage, attacks on humans, and impacts of human–wildlife conflict on both wildlife and people.

Profiling the Peer-reviewed Literature

There are several peer-reviewed articles on the illegal wildlife trade, wildlife trafficking, or components of those activities. Exploring how this scientific literature addresses principles of feminist political ecology, green criminology, and demographics helps baseline how the issue has been treated thus far, and it should inform future research or practice. Explicitly gendered research on wildlife trafficking is rare. Hübschle (2014) produced an in-depth overview of women's roles in wildlife trafficking in South Africa, which revealed a variety of ways that women participated in the illicit activity as professional hunters, drivers, safari operators, helicopter pilots, veterinarians, game rangers, smugglers, corrupt gatekeepers, law enforcement officials, customs and nature conservation officials, and money launderers. Supporting roles were also discussed, which delineated how Thai women living in South Africa were hired out of the Gauteng Province sex industry by poaching syndicates participated in pseudo-hunts so high-level offenders could avoid detection. Pires and Moreto (2016), Sollund (2017), and Zwier and Glajar (2018) published now highly cited review articles on the illegal wildlife trade and wildlife trafficking, yet the reviews did not incorporate gendered dimensions.

Cooney et al. (2017) explored the literature to determine trends in how local people engaged in community-based solutions designed to prevent wildlife poaching. Their analysis considered a range of poaching costs and benefits within the context of multiple sociodemographic factors, which included financial status and cultural values. Anti-poaching solutions were characterized by the degree to which they integrated local people's calculation of benefits and costs with demographics. The intersectionality of the variables was presented as an evaluative measure of community-based, anti-poaching programs; the benefits outweighed costs when all factors were considered. Although comprehensive in many ways, the framework did not consider gendered similarities and differences explicitly. This was unfortunate because men and women held different perceptions of wildlife poaching in certain contexts (e.g., Gore and Kahler, 2012).

The Convention on International Trade in Endangered Species of Wild Fauna and Flora (CITES) regulates legal trade in wildlife and, thus, is often used as a proxy for understanding illegal trade. Exploring the impact of CITES regulations on local communities and indigenous people is prudent, but unfortunately, most CITES-related studies do not incorporate gender into analysis in a meaningful way. For example, Persaud (2017) and Weber et al. (2015) have explored demand for high-value products, such as elephant (*Loxodonta africana*) ivory or polar bear (*Ursus maritimus*) hides, but did not consider gendered involvement, motivations, or activities beyond peripheral acknowledgment that the head of one ivory poaching syndicate was female (e.g., Queen of Ivory).

Case study-based articles offer peripheral mention of women who participated in wildlife trafficking activities, such as Gyrfalcon (*Falco rusticolus*) trafficking (Wyatt, 2011), but do not elaborate on or unpack the roles of women or the socioeconomic implications of participation. Ayling (2013) and Wittig (2016) published anecdotes of female Vietnamese and Thai sex workers being sent to South Africa to participate in rhino (*Diceros bicornis*) hunts and to traffic rhino horns to Southeast Asia, but they did not discuss the gendered dimensions of the convergent activities (e.g., victimization, recruitment). Price (2017) discussed the economic drivers of wildlife trafficking in Sub Saharan Africa but acknowledged only that the extant literature on wildlife crime was problematically gender blind (i.e., not considering gender as a factor in the topic at hand). Ramutsindela's (2016) profile of South Africa's Black Mamba all-female anti-poaching unit did not unpack the role of gender during analysis that connected the activities of this anti-poaching unit with security and terrorism. Cawthorn and Hoffman (2015) explored the intersectionality of wild meat consumption and food security in Equatorial Guinea, 2014–2015. They discussed women's preferences for romantic relationships with males who hunted for wild meat and their aversion to consume certain species with local taboos about infertility (e.g., palm civet [*Paradoxurus*], crowned guenon [*Cercopithecus pogonias*]).

Profiling the Gray Literature

The gray literature on wildlife trafficking and illegal wildlife trade is complementary to the peer-reviewed literature and offers many context-specific insights about the problem; it also profiles solutions. The UNODC's (2016) World Wildlife Crime Report was a seminal document that helped to benchmark the state of data on wildlife trafficking, report on trends, and forecast gaps. The role of women in wildlife trafficking was glaringly absent from the first report; women were identified as consumers of illegally traded wildlife products (e.g., pangolin scales used to make lactation tea). Global Financial Integrity's (2018) report on Illicit Financial Flows and the Illegal Trade in Great Apes surveyed financial actors across the illicit supply chain and discerned the most harmful dimensions of the activity on different great ape species. Some gendered dimensions of trade patterns were discussed; for example, the report noted that much of the illegal wild great ape meat was traded by women, who were often known as "market mamas," because law enforcement authorities were less likely to monitor females in markets compared with males. The report also identified similarities in how men and women used public transportation for moving illegal wild ape meat from forest to markets.

In Milliken and Shaw's (2012) The South Africa–Viet Nam Rhino Horn Trade Nexus report, the convergence of illegal sex work and wildlife trafficking was discussed; Vietnamese sex workers were exploited to traffic rhino horn using airplane carry-on luggage between countries. Associates of Vietnamese nationals who lived in South Africa were also exploited to purchase hunting licenses even though they were non-hunters. Finally, the gendered rhino horn consumption patterns were discussed, where Vietnamese women who were pregnant were advised not to take rhino horn medicinally; however, women did consume it to enhance their complexions.

The Center for International Forestry Research (CIFOR)'s (2019) report Towards a Sustainable Participatory and Inclusive Wild Meat Sector carefully considered the diverse roles that women played in wild meat hunting and trade. Whereas men were more likely to be the hunters across most ethnic groups studied, the report noted that for some ethnic groups (e.g., Aka Forest foragers) women were the primary hunters. Interestingly, the report profiled the differential impacts of wild meat consumption on women's income compared with men's income. Households that relied on men's income that was derived from hunting wildlife that was then sold to an intermediary and taken to market were not as stable as a household where a man sold wild meat directly in the market. Women's willingness to engage in alternative livelihood strategies that involved wildlife ranching was greater than men. The report noted that gendered data were challenging to obtain; records of women hunting were less frequent because women hunted opportunistically and for sustenance.

Minding the Gap

Clear gaps in knowledge remain with regards to women in wildlife trafficking. These gaps are problematic because by failing to account for women's participation in the harvest, transport, sale, and consumption of wildlife parts, decision-makers are deprived of essential information that could be used to develop solutions. Some of those solutions could be wildlife-focused, for example, how can illegal trade be mitigated more directly to reduce risks to endangered wildlife populations. Other solutions could focus more on people, for example, how can more inclusive, effective community-based conservation programs be designed, implemented, and evaluated. Moving forward, the scientific community and its partners should consider the roles of women more directly and the gendered dimensions of wildlife trafficking, particularly in conjunction with other sociodemographic factors, such as ethnicity and class. From a feminist political ecology perspective, all these factors inform an individual's knowledge, exit, and entry points in interacting with the environment that surrounds them. As culture and society evolve, so do the experiences of each member of the environment, and it is imperative that we account for every one of these experiences to facilitate the goal of diminishing wildlife crime.

Literature Cited

Agu, H. U., & Gore, M. L. (2020). Women in wildlife trafficking in Africa: A synthesis of literature. *Global Ecology and Conservation*, 23, e01166. https://doi.org/10.1016/j.gecco.2020.e01166.

Ayling, J. (2013). What sustains wildlife crime? Rhino horn trading and the resilience of criminal networks. *Journal of International Wildlife Law & Policy*, 16(1), 57–80. https://doi.org/10.1080/13880292.2013.764776.

Basu, G. (2013). The role of transnational smuggling operations in illicit supply chains. *Journal of Transportation Security*, 6(4), 315–328. https://doi.org/10.1007/s12198-013-0118-y.

Blaikie, P. (1985). *The Political Economy of Soil Erosion in Developing Countries*. Routledge. https://doi.org/10.4324/9781315637556.

Brisman, A. (2014). Of theory and meaning in green criminology. *International Journal for Crime, Justice and Social Democracy*, 3(2), 21–34. https://doi.org/10.5204/ijcjsd.v3i2.173.

Cao Ngoc, A., & Wyatt, T. (2013). A green criminological exploration of illegal wildlife trade in Vietnam. *Asian Journal of Criminology*, 8(2), 129–142. http://dx.doi.org.proxy2.cl.msu.edu/10.1007/s11417-012-9154-y.

Cawthorn, D.-M., & Hoffman, L. C. (2015). The bushmeat and food security nexus: A global account of the contributions, conundrums and ethical collisions. *Food Research International*, 76, 906–925. https://doi.org/10.1016/j.foodres.2015.03.025.

Clough, C., & Channing, M. (2018). Illicit financial flows and the illegal trade in great apes. Global Financial Integrity. Retrieved November 22, 2019, from https://gfintegrity.org/report/illicit-financial-flows-and-the-illegal-trade-in-great-apes/.

Coad, L., Fa, J. E., Abernethy, K., van Vliet, N., Santamaria, C., Wilkie, D. S., El Bizri, H. R., Ingram, D. J., Cawthorn, D.-M., & Nasi, R. (2019). Toward a sustainable, participatory and inclusive wild meat sector. Center for International Forestry Research (CIFOR). https://doi.org/10.17528/cifor/007046.

Cooney, R., Roe, D., Dublin, H., Phelps, J., Wilkie, D., Keane, A., Travers, H., Skinner, D., Challender, D. W. S., Allan, J. R., & Biggs, D. (2017). From poachers to protectors: Engaging local communities in solutions to illegal wildlife trade: Engage communities against illegal wildlife trade. *Conservation Letters*, 10(3), 367–374. https://doi.org/10.1111/conl.12294.

Cronin, D. T., Sesink Clee, P. R., Mitchell, M. W., Bocuma Meñe, D., Fernández, D., Riaco, C., Fero Meñe, M., Esara Echube, J. M., Hearn, G. W., & Gonder, M. K. (2017). Conservation strategies for understanding and combating the primate bushmeat trade on Bioko Island, Equatorial Guinea. *American Journal of Primatology*, 79(11), 22663. https://doi.org/10.1002/ajp.22663.

Eman, K., Mesko, G., Dobovsek, B., & Sotlar, A. (2013). Environmental crime and green criminology in South Eastern Europe—Practice and research. *Crime, Law and Social Change*, 59(3), 341–358. http://dx.doi.org.proxy1.cl.msu.edu/10.1007/s10611-013-9419-0.

Epanda, M. A., Mukam Fotsing, A. J., Bacha, T., Frynta, D., Lens, L., Tchouamo, I. R., & Jef, D. (2019). Linking local people's perception of wildlife and conservation to livelihood and poaching alleviation: A case study of the Dja biosphere reserve, Cameroon. *Acta Oecologica*, 97, 42–48. https://doi.org/10.1016/j.actao.2019.04.006.

Fa, J. E., Olivero, J., Farfán, M. A., Lewis, J., Yasuoka, H., Noss, A., Hattori, S., Hirai, M., Kamgaing, T. O. W., Carpaneto, G., Germi, F., Márquez, A. L., Duarte, J., Duda, R., Gallois, S., Riddell, M., & Nasi, R. (2016). Differences between pygmy and non-pygmy hunting in Congo Basin forests. *PLoS ONE*, 11(9), e0161703. https://doi.org/10.1371/journal.pone.0161703.

Fiorella, K. J., Camlin, C. S., Salmen, C. R., Omondi, R., Hickey, M. D., Omollo, D. O., Milner, E. M., Bukusi, E. A., Fernald, L. C. H., & Brashares, J. S. (2015). Transactional fish-for-sex relationships amid declining fish access in Kenya. *World Development*, 74, 323–332. https://doi.org/10.1016/j.worlddev.2015.05.015.

Gillespie, J., & Perry, N. (2019). Feminist political ecology and legal geography: A case study of the Tonle Sap protected wetlands of Cambodia. *Environment and Planning A: Economy and Space*, 51(5), 1089–1105. https://doi.org/10.1177/0308518X18809094.

Goetz, A. M. (2007). Political cleaners: Women as the new anti-corruption force? *Development and Change*, 38(1), 87–105. https://doi.org/10.1111/j.1467-7660.2007.00404.x.

Gore, M. L., & Kahler, J. S. (2012). Gendered risk perceptions associated with human-wildlife conflict: Implications for participatory conservation. *PLoS ONE*, 7(3), e32901. https://doi.org/10.1371/journal.pone.0032901.

Gore, M. L., Lute, M. L., Ratsimbazafy, J. H., & Rajaonson, A. (2016). Local perspectives on environmental insecurity and its influence on illegal biodiversity exploitation. *PLoS ONE*, 11(4). Opposing Viewpoints in Context. http://link.galegroup.com/apps/doc/A453448108/OVIC?u=msu_main&sid=OVIC&xid=c51ed93a.

Gore, M., Naess, L., Warinwa, F., Nyce, C., Teka, Y., & Andrew, C. (2019). Wildlife trafficking in Africa: Opportunities for science diplomacy. *Science & Diplomacy*.https://www.sciencediplomacy.org/article/2019/wildlife-trafficking-in-africa-opportunities-for-science-diplomacy.

Hoogenboom, B. (1991). Grey policing: A theoretical framework. *Policing and Society*, 2(1), 17–30. https://doi.org/10.1080/10439463.1991.9964629.

Howson, C. (2012). Women smuggling and the men who help them: Gender, corruption and illicit networks in Senegal. *Journal of Modern African Studies*, 50(3), 421–445. https://doi.org/10.1017/S0022278X12000183.

Hübschle, A. (2014). Of bogus hunters, queenpins and mules: The varied roles of women in transnational organized crime in Southern Africa. *Trends in Organized Crime*, 17(1–2), 31–51. http://dx.doi.org.proxy2.cl.msu.edu/10.1007/s12117-013-9202-8.

Leung, L., Miedema, S., Warner, X., & Homan, S. (2019). Making feminism count: Integrating feminist research principles in large-scale quantitative research on violence against women and girls. *Gender & Development*, 27(3), 427–447. https://doi.org/10.1080/13552074.2019.1668142.

Lowassa, A., Tadie, D., & Fischer, A. (2012). On the role of women in bushmeat hunting – Insights from Tanzania and Ethiopia. *Journal of Rural Studies*, 28(4), 622–630. https://doi.org/10.1016/j.jrurstud.2012.06.002.

Lynch, M. (1990). The greening of criminology: A perspective for the 1990s. *The Critical Criminologist*, 2, 11–12.

Mbete, R. A., Banga-Mboko, H., Racey, P., Mfoukou-Ntsakala, A., Nganga, I., Vermeulen, C., Doucet, J.-L., Hornick, J.-L., & Leroy, P. (2011). Household bushmeat consumption in Brazzaville, the Republic of the Congo. *Tropical Conservation Science*, 4(2), 187–202. https://doi.org/10.1177/194008291100400207.

Merchant, C. (1980). *The Death of Nature: Women, Ecology, and the Scientific Revolution* (Reprint edition). HarperOne.

Milliken, T., & Shaw, J. (2012). The South Africa-Viet Nam rhino horn trade nexus: A deadly combination of institutional lapses, corrupt wildlife industry professionals, and Asian crime syndicates (p. 173). TRAFFIC.

Mossoun, A., Pauly, M., Akoua-Koffi, C., Couacy-Hymann, E., Leendertz, S. A. J., Anoh, A. E., Gnoukpoho, A. H., Leendertz, F. H., & Schubert, G. (2015). Contact to non-human primates and risk factors for zoonotic disease emergence in the Taï region, Côte d'Ivoire. *EcoHealth*, 12(4), 580–591. https://doi.org/10.1007/s10393-015-1056-x.

Nightingale, A. (2006). The nature of gender: Work, gender, and environment. *Environment and Planning D: Society and Space*, 24(2), 165–185. https://doi.org/10.1068/d01k.

Nightingale, A. J. (2003). A feminist in the forest: Situated knowledges and mixing methods in natural resource management. https://www.era.lib.ed.ac.uk/handle/1842/1405.

Nightingale, A. J. (2011). Bounding difference: Intersectionality and the material production of gender, caste, class and environment in Nepal. *Geoforum*, 42(2), 153–162. https://doi.org/10.1016/j.geoforum.2010.03.004.

Nightingale, A. J., Bhattarai, A., Ojha, H. R., Sigdel, T. S., & Rankin, K. N. (2018). Fragmented public authority and state un/making in the 'new' republic of Nepal. *Modern Asian Studies*, 52(3), 849–882. http://dx.doi.org.proxy1.cl.msu.edu/10.1017/S0026749X16000500.

Nightingale, A., & Sharma, J. R. (2014). Conflict resilience among community forestry user groups: Experiences in Nepal. *Disasters*, 38(3), 517–539. https://doi.org/10.1111/disa.12056.

Nightingale, A. J., Bhattarai, A., Ojha, H. R., Sigdel, T. S., & Rankin, K. N. (2018). Fragmented public authority and state un/making in the 'new' republic of Nepal. *Modern Asian Studies*, 52(3), 849–882.

United Nations Office on Drugs and Crime (UNODC). (2016). World wildlife crime report: Trafficking in protected species, 2016. United Nations.

Ordaz-Németh, I., Arandjelovic, M., Boesch, L., Gatiso, T., Grimes, T., Kuehl, H. S., Lormie, M., Stephens, C., Tweh, C., & Junker, J. (2017). The socio-economic drivers of bushmeat consumption during the West African Ebola crisis. *PLOS Neglected Tropical Diseases*, 11(3), e0005450. https://doi.org/10.1371/journal.pntd.0005450.

Persaud, S. (2017). Losing our "CITES" on the "Traffic": How taxing ivory trafficking can save the African elephant from its bloody extinction. *Journal of International Wildlife Law & Policy*, 20(3–4), 295–320. https://doi.org/10.1080/13880292.2017.1403813.

Pires, S. F., & Moreto, W. D. (2016). *The Illegal Wildlife Trade* (Vol. 1). Oxford University Press. https://doi.org/10.1093/oxfordhb/9780199935383.013.161.

Pires, S. F., Schneider, J. L., & Herrera, M. (2016). Organized crime or crime that is organized? The parrot trade in the neotropics. *Trends in Organized Crime*, 19(1), 4. Academic OneFile.

Plumwood, V. (1993). *Feminism and the Mastery of Nature*. Routledge.

Poulsen, J. R., Clark, C. J., Mavah, G., & Elkan, P. W. (2009). Bushmeat supply and consumption in a tropical logging concession in northern Congo. *Conservation Biology*, 23(6), 1597–1608. https://doi.org/10.1111/j.1523-1739.2009.01251.x.

Price, R. (2017). Economic Drivers and Effects of the Illegal Wildlife Trade in Sub-Saharan Africa, 15.

Ramutsindela, M. (2016). Wildlife crime and state security in south(ern) Africa: An overview of developments. *Politikon*, 43(2), 159–171. https://doi.org/10.1080/02589346.2016.1201376.

Reuter, K. E., Randell, H., Wills, A. R., Janvier, T. E., Belalahy, T. R., & Sewall, B. J. (2016). Capture, movement, trade, and consumption of mammals in Madagascar. *PLoS ONE*, 11(2), e0150305. https://doi.org/10.1371/journal.pone.0150305.

Rocheleau, D. (1995). Maps, numbers, text, and context: Mixing methods in feminist political ecology. *The Professional Geographer*, 47(4), 458–466. https://doi.org/10.1111/j.0033-0124.1995.458_h.x.

Rocheleau, D. (Ed.). (1996). *Feminist Political Ecology: Global Issues and Local Experiences* (1 edition). Routledge.

Rocheleau, D. E. (2008). Political ecology in the key of policy: From chains of explanation to webs of relation. *Geoforum*, 39(2), 716–727. https://doi.org/10.1016/j.geoforum.2007.02.005.

Shiva, V. (1988). *Staying Alive: Women, Ecology, and Survival in India*. Kali for Women.

Sollund, R. (2017). Legal and illegal theriocide of trafficked animals. In J. Maher, H. Pierpoint, & P. Beirne (Eds.), *The Palgrave International Handbook of Animal Abuse Studies* (pp. 453–474). Palgrave Macmillan. https://doi.org/10.1057/978-1-137-43183-7_21.

South, N. (1998). A green field for criminology?: A proposal for a perspective— NIGEL SOUTH, 1998. *Theoretical Criminology*. http://journals.sagepub.com/doi/10.1177/1362480698002002004.

South, N., Brisman, A., & McClanahan, B. (2014). Green criminology. Obo. http://www.oxfordbibliographies.com/view/document/obo-9780195396607/obo-9780195396607-0161.xml.

Spira, C., Kirkby, A., Kujirakwinja, D., & Plumptre, A. J. (2019). The socio-economics of artisanal mining and bushmeat hunting around protected areas: Kahuzi–Biega National Park and Itombwe Nature Reserve, eastern Democratic Republic of Congo. *Oryx*, 53(1), 136–144. https://doi.org/10.1017/S003060531600171X.

Tagg, N., Maddison, N., Dupain, J., Mcgilchrist, L., Mouamfon, M., Mccabe, G., Ngo Badjeck, M. M., Tchouankep, M., Mbohli, D., Epanda, M. A., Ransom, C., & Fa, J. E. (2018). A zoo-led study of the great ape bushmeat commodity chain in Cameroon. *International Zoo Yearbook*, 52(1), 182–193. https://doi.org/10.1111/izy.12175.

US Census Bureau. (2016). Demographic analysis. The United States Census Bureau. https://www.census.gov/programs-surveys/popest/technical-documentation/research/demographic-analysis.html.

Weber, D. S., Mandler, T., Dyck, M., Van Coeverden De Groot, P. J., Lee, D. S., & Clark, D. A. (2015). Unexpected and undesired conservation outcomes of wildlife trade bans—An emerging problem for stakeholders? *Global Ecology and Conservation*, 3, 389–400. https://doi.org/10.1016/j.gecco.2015.01.006.

Wellsmith, M. (2011). Wildlife crime: The problems of enforcement. *European Journal on Criminal Policy and Research*, 17(2), 125–148. https://doi.org/10.1007/s10610-011-9140-4.

Wittig, T. (2016). IV. Poaching, wildlife trafficking and organised crime. *Whitehall Papers*, 86(1), 77–101. https://doi.org/10.1080/02681307.2016.1252127.

World Bank Group. (2016). Analysis of international funding to tackle illegal wildlife trade. World Bank. https://doi.org/10.1596/25340.

Wright, J., & Priston, N. (2010). Hunting and trapping in Lebialem division, Cameroon: Bushmeat harvesting practices and human reliance. *Endangered Species Research*, 11, 1–12. https://doi.org/10.3354/esr00244.

Wyatt, T. (2011). The illegal trade of raptors in the Russian Federation. *Contemporary Justice Review*, 14(2), 103–123. https://doi.org/10.1080/10282580.2011.565969.

Wyatt, T. (2013). *Wildlife Trafficking*. Palgrave Macmillan. https://doi.org/10.1057/9781137269249.

Zwier, P. J., & Glajar, D. S. (2018). Wildlife poaching and rule of law in Kenya. *Law and Development Review*, 11(2), 879–912. https://doi.org/10.1515/ldr-2018-0040.

Box 1

Voices from the Field

Offenders

Helen U. Agu and Meredith L. Gore

Offenders are one of the most important stakeholders in wildlife trafficking. They are considered the wrongdoers; *defenders* actively work to deter offenders from engaging in harmful behavior and sanction offenders when they violate rules. *Observers* may be an intentional part of an offenders' ingroup or professional network. Offenders may be pressured to engage in, and desist from, harmful activities by *influencers*. There can be associated socio-economic or cultural *benefits* for people who are associated with wildlife trafficking offenders. Some literature provides evidence that romantic connections are particularly important pathways for women into serious or transnational crime such as wildlife trafficking and illegal wildlife trade.[1] Conversely, unintended consequences or second- and third-order effects from wildlife trafficking may inflict *harm* on offenders or individual(s) in their networks. Deeper understanding about the motivations and intentions of offenders can advance insight about offender behavior, so that defenders may respond more effectively or prevent offending from occurring in the first place. Delineating types of offenders may inform targeted interventions or education and outreach. Women can be supporters (i.e., subordinate to the leading traffickers and, either under threat or 'voluntarily', executed orders of the leader or other members of the human trafficking networks), partners (i.e., women who have a relationship with a man and cooperate with him, in principle on the basis of equality, in conducting tasks and activities; such women may associate their illicit wildlife trafficking with intimate partner relationships), and madams (i.e., offenders who play a central role, lead criminal organizations, and coordinate human trafficking activities). There is a well-known sex bias in crime, and research on the role of women as perpetrators of crime has been limited. When we asked experts their perceptions of how, when, where, and why women engaged in wildlife trafficking (study design, methods, and analysis discussed in Agu et al. (2021)[2] some mentioned the fact that women are not often perceived as being offenders, as these voices from the field help illustrate:

> Women are very dangerous when they get into this business because they are not some of the obvious suspects…So, that is that quite a deceptive situation because when women get involved, they take very long to be

DOI: 10.4324/9781003121831-3

suspected of nobody associate them with smuggling nobody associates them with high level criminal activity in the word sector, which is seen as a man's operation.

(Female 1.3.1)

And this notion of a woman couldn't possibly be doing this is still very prominent. And it's and I think it is a big key missing part to ending illegal crime and, and trade because a lot of times what you don't realize is women will turn a blind eye just to not to get into trouble. Sometimes it's a very psychological thing. They don't want to meddle in male dominated issues, but turning a blind eye is kind of indirectly engaging with the sale, and they could end the crime itself. So, I have seen that women are often overlooked as potential offenders, which gives them the upper hand.

(Female 1.3.2)

Across borders, I've heard that women are less suspected and therefore they would be more, they would often be quite useful in actually being the trafficker instead of the one that normally carries things across the border.

(Male 12.9.1)

Although some experts have differing perceptions about the proportion of women offending in wildlife trafficking, some felt the role was still important. They shared with us, for example:

Women are hardly ever prosecuted for their crimes, and that's why it's difficult to quantify their real participation.

(Female 1.3.1)

So, there is clear evidence that women who may not be of African origin [are being caught in airports] with baubles concealed as necklaces or other fundamental items.

(Female 1.3.3)

Women play a very important role in all these different aspects because what they provide safe places for protest in the traffickers to hide in their homes (*sic*), when they are cutting out these operations, women will also be required small amount of money compared to men to move the products they can conceal their firewood or a market goods for them to be able to move products and so they operate at the lower levels of economic chain as far as traffic is concerned.

(Female 1.3)

Most of these exotic pets especially cheetahs and chimpanzees and orangutans, they are treated very badly, and most of them you know, most of them end up dying before they get sold. But when traffickers publicize these animals, they make them look so in need of care that women buy them just to take care of them without thinking they're promoting it.

(Female 1.5)

Voices from the field inspire interesting questions about the role of women as offenders in wildlife trafficking: Why women enter and desist from offending in wildlife trafficking? Do women enter on their own volition or are they coerced? Do crime patterns for women in wildlife trafficking differ; why or why not? What are the underlying motivations for different types of offending? Do wildlife trafficking offences overlap with other types of criminal offences and if so, under what conditions and why? Do theories and tools that help male offenders successfully desist from wildlife trafficking work as well with females? Why or why not? How long are the criminal careers of women in wildlife trafficking and how extensive is offending? Does women's liberation lead to an increase in female wildlife trafficking-related crime? Or, is men's wildlife trafficking-related crime decreasing while women's wildlife trafficking-related crime constant? Answers to these and other questions may help reduce harms associated with wildlife trafficking.

Notes

1 Jacqueline Hicks IDS, The role of gender in serious and organized/transnational crime 30 March 2021. https://opendocs.ids.ac.uk/opendocs/bitstream/handle/20.500.12413/16554/984_Gender_and_Serious_Organised_Crime.pdf?sequence=1.
2 The Michigan State University Human Subjects Protection Program approved the methods and analysis for the study (STUDY00003659) as exempt under 45 CFR 46.104(d) 2(ii).

Literature Cited

Agu, H.U., Andrew, C.A., and M.L. Gore. (2021). Mapping Terra Incognita: An Expert Elicitation of Women's Roles in Wildlife Trafficking. *Frontiers in Conservation Science*.

3 Women as Agents of Change in Efforts to Disrupt Illegal Wildlife Trade

Jessica Graham

Agents of Change and the Illegal Wildlife Trade

There is still high demand for effective strategies to disrupt illegal wildlife trade (IWT) and the corruption that fuels transnational environmental crime (Duffy, 2014; Elliott, 2007; Wyatt, 2013). Unfortunately, research on evaluating efficacy remains limited; extant literature suggests there is a relationship between gender and effective anti-poaching efforts, yet the nature of the relationship is not well understood (Agu and Gore, 2020; McElwee, 2012). Literature on women and conservation often discusses a need to include women as "assets" for conservation, and it promotes women's engagement in the social, political, ecological, cultural, and legal contexts of conservation (e.g., Gore and Kahler, 2012). This literature rarely discusses how or where to direct this inclusion. Thus, significant options remain to improve efforts that disrupt IWT supply chains.

This chapter leverages the knowledge base about gender from other sectors (e.g., Dollar et al., 2001) to help provide a foundation for future research and practice to combat IWT. By considering women as meaningful "agents of change," or individuals that may initiate, motivate, or manage shifts in the *status quo* (e.g., as policymakers, educators, community leaders, scientists, rangers), solution spaces can be identified, evaluated, and scaled to different IWT contexts. An "agents of change" perspective also helps to characterize problem spaces where women support the functioning of the IWT supply chain. In at least this regard, promoting discourse about women as agents of change in preventing IWT can inform a more gender-balanced approach for conservation, crime prevention, and crime reduction.

Agents of Change

Finding and supporting "agents of change" is a common objective in conservation and sustainability sectors because of the desire for shifting the *status quo* (e.g., Andriamihaja et al., 2021). Agents of change are different from assets because they are positioned within their social network in such a way to engender transformation. The human actor model considers an agent of change to be a function of the individual's action. Action is the sum of meanings (i.e., motivations to engage in a particular behavior), means (i.e., resources held and needed to behave in a particular manner), and activity (i.e., interactions with other individuals that occur during the behavior) (Wiesmann et al., 2011). Becoming and remaining an agent of change in the context of IWT requires a woman to adjust and to mobilize her meaning, means, and activity in various situations (Andriamihaja et al., 2021). Conducting an "agency analysis" is a common methodology for identifying agents of change; however, the lack of information about women in wildlife trafficking first requires a review of the literature to set a baseline upon which to conduct research. This chapter contributes such a review to help identify where women serve currently, or may serve, as agents of change in IWT.

A Brief Review of Poaching's Past and Present

The deep historical linkages between humans and wildlife are often framed according to the roles that men and women played in securing food. For example, the "hunter-gatherer theory" has long posited prototypical differences between males and females (Silverman et al., 2007). Successful gathering involved females seeking out and locating edible plants regardless of the "diverse configurations of vegetation" or season—a "landmark" strategy. Male's generalized ability to use an "orientation" strategy in navigation helped them to hunt successfully (ibid). This type of primitive spatial duality can be interpreted within the present day, where cultural narratives often frame males as being the hunters and traders of wildlife, such as jaguars (*Panthera onca*) or saiga (*Saiga tatarica*), because of the dangers associated with hunting; hunting by males promoted confrontation with potentially dangerous animals or long-distance travel over challenging terrain. Females were associated with fuelwood collection, coastal and near-community subsistence fishing, non-timber forest product production, and carrying goods on their final leg of the journey to market (McElwee, 2012).

In the 18th century, Europeans began to poach in earnest as poor members of society who were disadvantaged structurally engaged in alternative methods to source food for survival (Umar, 2017). Nature was viewed as an obstacle in the progress of civilization, and it was treated like a commodity (Moyle and Orland, 2004). This "tradition" was perpetuated as patriarchal Europeans engaged in colonization and exploitation of wildlife around the world. Overexploitation of wildlife by colonizers was not restricted institutionally, and conservation was relegated to a moral precept. Between 1600 and 1900, some 75 mammal species became extinct, which was due largely to the direct consequence of colonial exploitation of nature (Umar, 2017). The depletion and sharp decline of wildlife was due to both an agrarian and early industrial society, which was essentialized in part through the hunter-gatherer theory.

Poaching and hunting were regulated eventually by feudal landowners, and a new criminal industry was born. Today, environmental crime, which includes wildlife trafficking, is one of the top four most lucrative forms of transnational organized crime, behind money laundering, narcotics trafficking, and counterfeit products. Although difficult to estimate, wildlife trafficking is valued conservatively at 20 billion USD each year (UNODC, 2016). Although not all IWT is organized and criminal, some of it may be opportunistic across different parts of the supply chain. The IWT that drives species to critically low population levels is driven by demand and sustained by greed.

Women in the Supply Chain of the Illegal Wildlife Trade

Both legal and illegal supply chains involve flows of products, services, information, and finances from a source-to-end user. Different actors support these flows in unique ways, such as sourcing, manufacturing, trafficking, and selling products to consumers in destination markets. The United Nations Office on Drugs and Crime (UNODC) models six links or stages in illegal supply chains: poachers, runners/brokers, intermediaries, exporters, importers/wholesalers, and retail traders (UNODC, 2020). The proportion of males and females involved in each stage of the IWT supply chain is not known completely because of the lack of data that are collected on the issue; generally, women are not assets along the IWT supply chain and, if they are, it is rarely in a leadership role. When women are identified as being assets, they are associated most commonly with production and consumption links. The six links of the IWT supply chain complement the six roles that women may assume in wildlife trafficking: offenders, defenders, observers, influencers, beneficiaries, or persons harmed (Agu and Gore, 2020).

The IWT "begins" with poaching by individuals who derived their livelihood primarily from natural resources; poaching and poachers disproportionately occurred in countries where wildlife live, and male protagonists played the leadership role in hunting for the IWT. However, in source locations, anti-poaching laws and programs inflicted significant burdens on household women because they led household activities, such as fuelwood collection, subsistence hunting, and fishing. These women were forced to choose between violating rules or seeking alternatives, such as having to travel to farther locations to fulfill their household needs. Household responsibilities also affected women's attitudes about poaching, but scant literature exists to equivocate trends (e.g., Sundström et al., 2019). Lowassa et al. (2012) found that female participants in their study area were supportive of males who poached for food or income.

Poached wildlife was trafficked through intermediaries who navigated transit across national, sub-regional, and regional geographies. Trafficking at this stage often involved bribes to law enforcement authorities and government officials (e.g., turn a blind eye, forged paperwork, safe passage). Males were more likely to obtain employment as a government ranger, official, inspector, or law enforcement official, which resulted in gender interaction norms (i.e., expectations of what is traditionally masculine or feminine behavior) that underlay successful corrupt relationships and kept women from participating in this stage of the IWT (McElwee, 2012). Successful intermediaries had international professional networks or contacts with organized crime. Products travelled onward toward destination markets where wholesalers and retailers sold products to consumers who used them in diverse locations (e.g., home, restaurant, life cycle event, business) (UNODC, 2020). McElwee (2012) found that the gender of intermediaries was often not recorded by authorities who made arrests and seizures. Occasional media reports of seizures noted that women were arrested for moving illegal wildlife products into cities from rural areas and then onward for future transit.

The well-publicized wildlife trafficking case of the infamous "Ivory Queen," Yang Feng Glan, influenced public perceptions of the role of women in wildlife trafficking. Yang, who was a Chinese National based in Tanzania, operated globally and trafficked wildlife across multiple continents before her arrest in 2019. She was charged with smuggling USD $5.6 million worth of ivory from more than 350 elephants (*Loxodonta africana*) and was sentenced to 15 years in prison (Reuters, 2019). Although this high-profile case was known widely, empirical evidence was lacking and, ultimately, trends associated with these assets cannot be determined at this time.

Data about the end-stages of the IWT supply chain is the most detailed in terms of gender. Affluent businessmen were the main buyers of illegally traded ivory in mainland China, for example, which challenged the common notion that buyers were predominantly elders who kept with traditional norms of ivory carvings as cultural symbols (Venkataraman, 2007). Males were identified commonly as consumers of illegal wild meat for use in luxury dining (e.g., exotic delicacy) and wellness (e.g., male virility). The socio-cultural experience of business dining was important for men, and consumption of illegal wildlife products coupled with live entertainment was a symbol of affluence and status (Venkataraman, 2007). Women were featured as majority purchasers of illegal wildlife products for ornamental and decorative use (e.g., Margulies et al., 2019).

The extent that women served as assets along the IWT supply chain remains mostly unknown. Practically, we know that women lack the same opportunities for access and resources needed to participate in IWT, in general, and are not able to compete on an equal footing with their male counterparts, specifically. Thinking about women as agents of change in the IWT supply chain is also similarly under-researched; however, baselining such information could have broad practical implications for efforts to reduce risks associated with IWT. Characterizing how, when, and where women served as agents of change in the IWT would create entry points for lesson learning, scaling interventions, and allocating resources.

Reviewing the Literature on Women as Agents of Change

Multiple search engines were used to identify relevant publications on IWT and women, which included EBSCO Information Service and Google Scholar. Identical keyword searches were conducted using: wildlife crime, poaching, women in wildlife, women in conservation, women and anti-poaching, and wildlife trafficking and women. Reference citation tracking was used to cross-check the search where dozens of documents were identified. Sources were analyzed according to the three dimensions of "agents of change" and six roles for women in wildlife trafficking. Sources were scanned, ordered, reviewed, and compared for any roles that women played as agents of change (LeCompte and Goetz, 2018). Four roles exemplified women as agents of change: conservation scientist–activist, policy drivers, leadership/management, and rangers.

Women as Conservation Scientist–Activists

Literature that tracked the history of conservation science acknowledged a number of scientist–activists who not only asked, and answered, challenging scientific questions about conservation, but also changed the gendered *status quo* that was associated with fieldwork and conceptualizations of animal behavior. In helping to advance the political landscape of conservation in the late 19th and early 20th centuries, many women as conservation scientist–activists crossed academic boundaries and social norms (Mahoney, 2019).

Primatologists, such as Jane Goodall and Dian Fossey, both white women from the north who were also trained at universities based in the global north, traveled, respectively, to study chimpanzees in Tanzania and gorillas in Rwanda. Their fieldwork protocols and procedures are known widely, and their roles as agents of change in conservation, both positive and negative, have been chronicled extensively in international contexts. Their characteristics as scientists (e.g., age, sex, marital status, number of children) are as well-known as their science (e.g., epistemology, ontology). Marine biologist Rachel Carson, also a white woman trained in the global north, was an elegant writer, whose "Silent Spring" helped usher in the modern environmental movement in the United States. It is not a coincidence that these renowned female scientists advanced professionally during the second wave of feminism, environmentalism, and on-the-ground animal rights activism (Gaarder, 2011), which included other famous women, such as Vandana Shiva from India. These women engaged in repeated, small-scale, and context-laden resistance that helped to advance public conversations about conservation, animal rights, and women in science. Their public efforts helped to raise awareness and to promote changes within their institutions (e.g., academia, conservation). In many ways, the modern environmental defenders movement, which does not always focus on conservation *per se*, yet is embodied by many female agents of change, represents more contemporary women as conservation scientist–activists. Well-known defenders, such as Berta Cáceres (Honduras), Wangari Maathai (Kenya), and Sheila Watt-Cloutier (Canadian First Nation), positioned themselves similarly as voices that reached between the local and the global and, thus, illustrate women as agents of change.

Women as Policy Drivers

As women helped to advance change in conservation science–activist spaces, so too did they help to advance new legislation about wildlife poaching, hunting, and animal rights (e.g., Birke, 2000), often in direct response to degraded natural resources or wildlife populations. Some scholars linked the advancement of women's rights and equality that was associated with second wave feminism with the animal rights legislation movements,

> They were...opposed to all injustice, including human mistreatment of animals. Feminism was...part of a wider set of problems; animal cruelty reflected a greater barbarism leading to mistreatment of humans. Accordingly, they actively campaigned for legislation to protect animals and the environment and lobbied the fledgling United Nations to include such measures.
>
> (pp. 693–719)

Even before the environmental movement took hold in the United States, women served as under-appreciated policy drivers in conservation. For example, U.S. President Theodore Roosevelt is widely known to have supported hunting regulations to advance conservation of wildlife. Historians note that his second wife was highly influential in encouraging his broad perspective on conservation, which linked ethical hunting practices and land preservation for wildlife during his presidency (Gould, 2014). Bethine Church, wife of U.S. Senator Frank Church of Idaho, was known to be highly influential in supporting his sponsorship of the United States. The Wilderness Act of 1964 and other wilderness-related laws created processes for designating wilderness that included civic engagement (Dant, 2008). Contemporary female regulators have driven legislative change to help solve environmental challenges, such as IWT. In the United States, the League of Conservation Voters found that between 2006 and 2018, the average voting scorecard for women senators was 71 compared with 46 for their male counterparts (Women's Media Center, 2019). In the 117th U.S. Congress, major bills were introduced by women, which included Representative Debbie Dingell from Michigan, who introduced the Recovering America's Wildlife Act of 2021, with 85 cosponsors. H.R. 4182 was introduced by Representative Marie Newman from Illinois, to "require the President to declare the global wildlife extinction crisis a national emergency."

Women in Leadership and Management

The presence of women in leadership and management positions is not always discussed explicitly in the literature, although it can be implicit in authorship by-lines and organizational charts. The women who served as lead authors for reports or publications, in management positions, and who were tasked with the responsibility to achieve organizational missions, strategically planned and filled employee pipelines.

Beyond discussions about the implications of gender on publishing rates and outlets (e.g., Primack et al., 2009), female authors set the agenda or framed research articles, popular press stories, or government reports. In this regard, women used their voice to tell people what to *think about* and not *what to think*; for example, women helped to mainstream conversations about equity, gender-based violence, animal welfare, or inclusion of indigenous voices in conservation (Ertürk, 2004). The National Geographic Magazine published extensively on the IWT; female journalists authored investigations that were read widely, such as those about pangolin (*Pholidota*) trafficking for traditional medicine, the captive lion industry (*Panthera leo*), and hyacinth macaw (*Anodorhynchus hyacinthinus*) egg smuggling. These stories often asked challenging questions and built an understanding of the issues at hand. Importantly, many female authors raised awareness with their writings, and they used their platforms to laud efforts on successful IWT reduction efforts. In at least these regards, these individuals served as agents of change.

Many formal or professional leaders and managers in wildlife trafficking-oriented nongovernmental organizations were women. These organizations have released high-impact reports and associated briefings that contributed key insights for policymakers, donors, and scientists. For example, the World Wildlife Crimes Report, which was developed by the UNODC, serves as a key resource for the conservation community and provides a glimpse of global illegal wildlife trafficking trends. Fifty-eight percent of the acknowledged contributors to the 2020 report were women (ibid.). In the World Wide Fund for Nature's 2020 Living Planet Report, nearly 50% of the authors were women (WWF, 2020).

Indigenous female leaders have mobilized new networks of wildlife crime observers where none existed before. For example, Team Lioness has engaged women in multiple Maasai sub-tribes in Kenya, in collaboration with female leaders from the International Fund for Animal Welfare (IFAW). One female leader from IFAW noted, "We are finding that where the women have trusted channels of other females to report to, they do. As a result, we have been able to save the lives of a pride of lions and 2 elephants" (Forbes, 2019).

Women as Rangers

Multiple policies and programs designed to reduce IWT included building capacity and efficacy of "boots on the ground." One of the myriad implications of this approach was that conservation rangers served as a force multiplier against all types of organized crime, some of which were linked to the IWT. Wildlife rangers and anti-poaching units were on the "front lines" of the "fight" against IWT. Rangers often played a critical role in protecting 30% of the planet's ecosystem, which covered remote areas and ecosystems and wildlife that they protected directly (IRF, 2019). In some instances, wildlife rangers were asked to "flex" at work, shifting duties to help fight forest fires, such as the 2020 blazes in Australia. Over 7,000 wildlife rangers from 28 countries summarized many of their professional challenges, which included being underpaid, uninsured, and inadequately trained, working temporary contracts, lacking basic equipment that included clean drinking water, boots, and raincoats, and inadequate accommodation, while spending weeks away from home (WWF, 2019).

At the 2019 International Ranger Federation World Congress in Chitwan, Nepal, more than 550 rangers from 70 countries agreed upon the Chitwan Declaration, which detailed the needs and priorities for rangers to do their work more professionally, safely, and responsibly, and it highlighted the need for gender equality in hiring, pay, and promotion within the workforce (IRF, 2019). Not only was there no gender equality within the role of rangers globally, but just as importantly, there was a need for *gender equity* within ranger workforces. Typically, women who occupied positions within the ranger workforce served in administrative roles with little ability for professional development, promotion, or leadership positions.

The latest assessment estimated that 7.5% of the world's total ranger workforce were women (WWF, 2019). Although this was a small percentage of the total population, it demonstrated a clear opportunity for women as agents of change because anti-poaching units and ranger roles have been occupied by men traditionally. Female rangers faced unique and additional hardships individually and as part of ranger units, such as not being integrated sufficiently into a team, lack of privacy and lodging (WWF TNRC, 2020), and a lack of options for disposing of feminine waste products. Female rangers faced higher levels of harassment than men and also faced gender-based violence from fellow rangers and supervisors in the field and in the office (Seager, 2020).

When female rangers were able to engage with communities as part of their foot patrols, compared with serving in an administrative capacity, they demonstrated repeatedly a strong ability to de-escalate conflicts; this was a particularly important skill when rangers were confronted by poachers in the bush. Female rangers often used engagement as a first step toward de-escalation, whereas men were socialized to use enforcement as an initial step. Gender-interaction norms were also an important dimension of foot patrols; female rangers were able to conduct home and body searches on other women where they knew women often hid contraband (Seager, 2020).

Female rangers built different relationships with communities than did their male counterparts. Due to cultural barriers, male rangers were often unable to form relationships with women in communities. Female rangers had an ability to connect with local community youth, which resulted in better two-way communication, raised awareness about conservation, and increased information flow between rangers and communities to learn about important wildlife trafficking trends, poachers, and other illegal wildlife activities. Additionally, given gender interaction norms and traditional roles as educators and primary caretakers of children across regions that were impacted heavily by illegal wildlife trafficking, female rangers discouraged future generations from poaching more effectively than did men.

Given the tremendous potential of women to serve as agents of change for effective ranger roles, and perhaps in response to the logistical challenges of integrating women into all-male ranger units, all-female ranger units are being deployed widely. Each unit-model differs by location and context, yet all devote significant attention to rigorous training, selection, and recruitment that translates into effective ranger outcomes, low turnover, and zero corruption within the units. All-female ranger units are associated with resurgent wildlife activity, decreased poaching rates, and enhanced community security. These types of units have illustrated a strong ability to strengthen community, to contribute to economic opportunity, and to inspire youth, especially young women, to pursue a profession in conservation. One female ranger even arrested her husband for poaching activities (Graham, 2020)! Profiling all-female, anti-poaching units illustrates the nature of women individually and collectively who serve as agents of change.

The Akashinga in Zimbabwe. Akashinga, which means "brave one" in Shona, was established in 2017 in Zimbabwe's Lower Zambezi Valley, where one of the largest remaining contiguous savanna elephant populations resides. A rigorous interview and physical fitness selection program was developed by the International Anti-Poaching Foundation (IAPF) to build ranger capacity and also to provide new opportunities for the most marginalized and vulnerable members of the community. The initial cadre of women who passed the grueling endurance test were all survivors of sexual assault and/or domestic violence. They were AIDS orphans, single mothers, and abandoned wives. Of 189 men who had conducted the same selection training six years earlier by the IAPF, three voluntarily withdrew in 24 hours. Of the 37 resilient women who went through the same training years later, three voluntarily withdrew after a period of 72 hours.

Since 2017 when the program began, IAPF has utilized a paramilitary-like training model for the Akashinga. Female rangers are armed and outfitted like male military troops, which include grooming standards, uniforms, and appearance. The program includes education, food, and nutrition programs for youth that are administered by the female rangers in their communities, which generates strong stakeholder support from local tribal elders. The female ranger position is highly respected and sought after within the community with more younger women who aspire to become rangers. They now have over 150 rangers who have conducted nearly 200 arrests and contributed to the country's 80% reduction in poaching rate.

A similar model was established in Kenya's Segera Conservancy in 2020. This unit came about because women from other areas were collecting firewood for fuel illegally, and the existing male rangers did not want to arrest women, so there was a need for female rangers to be brought in to carry out the arrests. This unit is still in its infancy, but it is unique in its training approach because they integrate yoga and meditation into the rigorous six-month ranger training program for women. After training, they join the male rangers in integrated units.

The Black Mambas in South Africa. In 2013, the Black Mambas was established in Balule Nature Reserve, South Africa, which is adjacent to Kruger National Park. The Black Mambas have adopted techniques from British Police; they dress alike and look the part of a ranger in a visual deterrent for crime prevention. The Black Mambas are unarmed to avoid an arms race with poachers, which helps rangers avoid fatal confrontations. Female rangers patrol the periphery of poaching zones to help collect and to disseminate intelligence into a centralized control room. They embark on 21-day patrols, which can be a challenge for women with families and small children. Recruits undergo a rigorous three-month training that incorporates training in physical and classroom skills, such as surveillance practices. Recruits must also complete a detailed interview and fitness test and pass a polygraph exam like men. A 2020 impact survey conducted on the Black Mambas found that female rangers self-reported dramatic improvements in their self-esteem and gender role expectations. Additionally, the Black Mambas have had a positive influence on conservation awareness, which led to a decrease in support for poaching activities in surrounding communities (Danoff-Burg & Ocanas 2020). Approximately 30 women have been recruited into service to date and contributed to a reduction in poaching and snaring events in the areas that they patrol by 76%, which far exceeded what all-male units typically accomplished in a similar-sized area.

Team Lioness in Kenya. In 2017, the IFAW supported the creation of Team Lioness in Kenya. Under the oversight of the Olgulului Community Wildlife Rangers, Team Lioness protects Maasai communities that surround Amboseli National Park. Maasai women are deeply connected to their communities and land, making them key agents of change for local conservation efforts with insider knowledge and unique perspectives to offer counterpart male ranger units. Team Lioness rangers are part of a larger, mixed gender model, with one female representative coming from each of the eight sub-clans of the Maasai Indigenous Community to ensure greater community buy-in and fuller representation among the existing male units. When women rangers are recruited and trained, they are with the men for the first 90 days, and value is placed on integrating the teams early for cohesion. Men are asked to learn more non-violent and non-aggressive tactics from the women, and women are asked to push themselves in their physical fitness training alongside the men. The Team Lioness recruitment selection process is rigorous and incorporates the tribal leaders' support; selection is based on leadership, academic achievements, physical fitness, and integrity. Team Lioness does not have a minimum educational requirement. Many of the younger women on Team Lioness are among the first in their families to secure employment. Therefore, the ranger position is considered an extremely high honor, which likely also contributes to the greater scrutiny and importance placed on ensuring zero corruption.

Asia's All-Female Units. Two all-female anti-poaching units in Asia further illustrate female rangers as agents of change: a small group in Nepal and the Lion Queens of India. Thanks in part to these female units, Nepal's protected areas achieved near-zero poaching of rhinos (*Rhinoceros unicornis*) during 2011–2018. Since 2007, the Lion Queens unit that patrolled and protected the Gir Forests in Gujarat State has grown to >50 women. Although the Lion Queen's salary is below their male counterparts, their dedication and commitment to saving lions, leopards (*Panthera pardus*), and other wildlife is widely known. Despite the long hours, the Lion Queens have arrested dozens of poachers while supporting their family household income concomitantly (Nayar, 2016).

From Subjects to Agents of Change

Women have served as active agents of change in IWT in multiple ways, such as conservation scientist–activists, policymakers, managers and leaders, and rangers. In many instances, women exhibited effective leadership and conflict resolution styles that were engaging, communicative, and inclusive, which benefited sustainable wildlife conservation. Female rangers were often perceived as more honest and responsive than their male counterparts, which was an important characteristic in countries where corruption was known to run rampant in the conservation sector (Sundström and Wyatt, 2017). Although the temporal impacts of women who served as agents of change are still mostly unknown, they can be measured and included as part of strategic or financial planning.

The possibilities of women as agents of change in the ranger sector are noteworthy, in part, because of a ranger's unique position in society. The requirements to work closely with local communities and to raise awareness about conservation within the community are nothing new to any ranger, but female rangers exhibit a differential ability to serve as educators and communicators within communities for the benefits of conservation. Gender equality among ranger workforces has the potential to change women's influence within mixed gender units and to help change expectations of standards for professionalization. This creates the opportunity for more all-female, anti-poaching units to expand beyond the few countries in which they exist currently to influence future generations of young women and to serve as role models in the conservation effort. Additional research on this topic would be widely beneficial across conservation contexts.

This chapter begins to help address how, when, and why women are agents of change in IWT. Across the recurring themes in the literature, women's central status in communication networks (informal, familial, often rural) positions them for intelligence and information-sharing, which is critical for crime prevention. Second- and third-order effects emerge when female rangers are at the front lines and are given equity and equality, such as overall organizational performance, effectiveness, and authentic leadership (e.g., Lopez et al., 2015). Placing women in a non-traditional position as a ranger inspires community youth and results in generational impacts. Although there has been much attention placed on the issue of IWT in Africa, there is a continued need for further research and advocacy for women in conservation practices among other regions around the globe, such as Latin America, the Middle East, and Southeast Asia. For example, to date, there are no known all-female ranger units in Latin America or South America. This is a key area for additional research in which to explore the roles of women in IWT. There remains an untapped potential to provide more meaningful roles for women as agents of change in IWT.

Literature Cited

Agu, H. U., & Gore, M. L. (2020). Women in wildlife trafficking in Africa: A synthesis of literature. *Global Ecology and Conservation*, 23, e01166.

Andriamihaja, O. R., Metz, F., Zaehringer, J. G., Fischer, M., & Messerli, P. (2021). Identifying agents of change for sustainable land governance. *Land Use Policy*, 100, 104882.

Birke, L. (2000). Supporting the underdog: Feminism, animal rights and citizenship in the work of Alice Morgan Wright and Edith Goode. *Women's History Review*, 9(4), 693–719.

Dant, S. (2008). Making wilderness work: Frank Church and the American Wilderness Movement. *Pacific Historical Review*, 77(2), 237–272.

Danoff-Burg, D., Ocanas, A. (2020). 'Individual & Community-Level Impacts of the Black Mamba Anti-Poaching Unit', The Living Desert Zoo and Gardens, pp. 1–49. (Confidential Copy)

Dollar, D., Fisman, R., & Gatti, R. (2001). Are women really the "fairer" sex? Corruption and women in government. *Journal of Economic Behavior & Organization*, 46(4), 423–429.

Duffy, R. (2014). Waging a war to save biodiversity: the rise of militarized conservation. *International Affairs*, 90(4), 819–834.

Chiu, B. (2019). Indigenous women leading charge in wildlife conservation. Forbes. com (Accessed December 1, 2021).

Elliott, L. (2007). Transnational environmental crime in the Asia Pacific: An 'un(der) securitized' security problem? *The Pacific Review*, 20(4), 499–522.

Ertürk, Y. (2004). Considering the role of men in gender agenda setting: Conceptual and policy issues. *Feminist Review*, 78(1), 3–21.

Gaarder, E. (2011). Where the boys aren't: The predominance of women in animal rights activism. *Feminist Formations, 23*(2), 54–76.

Graham, J. (2020). Unbreakable: Females fighting poaching. *TNRC Publications.* To be issued in Fall of 2020.

Gore, M. L., & Kahler, J. S. (2012). Gendered risk perceptions associated with human wildlife conflict: Implications for participatory conservation. *PLoS ONE, 7*(3), e32901.

Gould, L. L. (2014). *American First Ladies: Their Lives and Their Legacy.* Routledge.

International Ranger Foundation (IRF). IRF [International Ranger Federation]. 2019. Who is a ranger? https://www.internationalrangers.org/

LeCompte, M. D., & Goetz, J. P. (1983). Annual meeting of the American Educational Research Association. In *Playing with Ideas: Analysis of Qualitative Data.* Montreal.

Lopez, C. G.-G., Alonso, F. M., Morales, M. M., & Leon, J. A. M. (2015). Authentic leadership, group cohesion and group identification in security and emergency teams. *Psicothema, 27*(1), 59–64.

Lowassa, A., Tadie, D., & Fischer, A. (2012). On the role of women in bushmeat hunting – Insights from Tanzania and Ethiopia. *Journal of Rural Studies, 28*(4), 622–630.

Mahoney, S. (2019). *Volume 4: The Women's Issue, Women, Hunting, & Conservation.* The Modern Huntsman.

Margulies, J. D., Wong, R. W. Y., & Duffy, R. (2019). The imaginary 'Asian super consumer': A critique of demand reduction campaigns for the illegal wildlife trade. *Geoforum, 107,* 216–219.

McElwee, P. D. (2012). *The Gender Dimensions of the Illegal Trade in Wildlife: Local and Global Connections in Vietnam, in Gender and Sustainability: Lessons from Asia and Latin America.* The University of Arizona Press, 71–93.

Moyle, P. & Orland, M.A. (2004). Chapter 2: A history of wildlife in North America. In Moyle, P. & Kelt, D (Eds.) *Essays on Wildlife Conservation.* MarineBio, University of California, Davis. https://www.marinebio.org/creatures/essays-on-wildlife-conservation/3/

Nayar, A. (2016). Why these fearless women are being called the "lion queens" of India. Huffpost.com. https://www.huffpost.com/archive/in/entry/gir_0_n_9288528

Reuters. (2019, February 19). Tanzania court convicts 'ivory queen' for trafficking elephant tusks. *New York Times.* https://www.nytimes.com/2019/02/19/world/africa/tanzania-ivory-queen-convicted.html.

Seager, J. (2020). *Gender Balance in the Ranger Workforce.* Bentley University.

Silverman, I., Choi, J., & Peters, M. (2007). The hunter-gatherer theory of sex differences in spatial abilities: Data from 40 countries. *Archives of Sexual Behavior, 36*(2), 261–268.

Sundström, A., Linell, A., Ntuli, H., Sjöstedt, M., & Gore, M. L. (2019). Gender differences in poaching attitudes: Insights from communities in Mozambique, South Africa, and Zimbabwe living near the great Limpopo. *Conservation Letters, 13*(1), 1–8.

Sundström, A., & Wyatt, T. (2017). Corruption and organized crime in conservation. *Conservation Criminology,* 97–113. https://doi.org/10.1002/9781119376866.ch6.

Primack, R. B., Ellwood, E., Miller-Rushing, A. J., Marrs, R., & Mulligan, A. (2009). Do gender, nationality, or academic age affect review decisions? An analysis of submissions to the journal Biological Conservation. *Biological Conservation, 142*(11), 2415–2418.

Umar, M. (2017). Mapping the history of illegal wildlife trade and construction. *Bharati Law Review*, 145–165.

UNODC. (2016). *World Wildlife Crime Report, Trafficking in Protected Species.* United Nations.

UNODC. (2020). *World Wildlife Crime Report, Trafficking in Protected Species.* United Nations.

Venkataraman, B. (2007). *A Matter of Attitude: The Consumption of Wild Animal Products in Hanoi.* Hanoi: TRAFFIC Southeast Asia, Greater Mekong Programme.

Wiesmann, U. M., Ott, C., Ifejika Speranza, Wiesmann, Urs; Ott, Cordula; Ifejika Speranza, Chinwe; Kiteme, Boniface; Müller-Böker, Ulrike; Messerli, Peter; Zinsstag, Jakob (2011). *A human actor model as a conceptual orientation in interdisciplinary research for sustainable development.* In: Wiesmann, Urs Martin; Hurni, Hans (eds.) Research for Sustainable Development: Foundations, Experiences, and Perspectives. Perspectives of the Swiss National Centre of Competence in Research (NCCR) North-South: Vol. 6 (pp. 231-256). Bern: Geographica Bernensia.

Wilderness Society. (n.d.). 11 Women Who Made Wilderness History. Accessed: https://www.wilderness.org/articles/article/11-women-who-made-wilderness-history.

Women's Media Center. (2019). https://womensmediacenter.com/news-features/women-in-congress-get-higher-scores-on-environmental-issues.

World Wildlife Fund. (2019a). *Life on the Frontline, A Global Survey of the Working Conditions of Rangers.* World Wildlife Fund.

World Wildlife; Targeting Natural Resource Corruption Project. (2020). *Female Rangers and Anti-Poaching Strategies to Stem Corruption.* Targeting Natural Resource Corruption Project. https://www.worldwildlife.org/pages/tnrc-blog-female-rangers-and-anti-poaching-strategies-to-stem-corruption

World Wildlife Fund. (2020). *WWF Living Planet Report 2020.* World Wildlife Fund.

Wyatt, T. (2013). The security implications of the illegal wildlife trade. *The Journal of Social Criminology*, Autumn/Winter 2013, 130–158.

4 Using a Feminist and Green Social Justice Perspective to Better Understand Governance of Wildlife Trafficking in Mexico

Inés Arroyo-Quiroz, Jesús Ignacio Castro Salazar, and Serena Eréndira Serrano Oswald

Wildlife trafficking can involve multiple serious aspects of harm, criminality, violence, and violations of the rule of law, as recognized by international legal frameworks such as the United Nations Convention against Transnational Organized Crime (UNTOC) or the United Nations Convention Against Corruption (UNCAC). Any country may be a source, a location for transit, and/or destination country for wildlife trafficking. Thus, all countries share a role in helping to mitigate these crimes (UNODC, 2016; 2020). Mexico is a uniquely important country for wildlife trafficking due to its high levels of biodiversity, geographical centrality in North America, and the purchasing power of its elites (Barth, 2017). Today, Mexico's major role in global wildlife trafficking is as an entrepôt nation; the country continues to report high levels of import, smuggling, and re-export of non-native species (e.g., reptile skins for leather). There have also been high levels of trafficking of native species, particularly reptile skins, psittacines and other birds, and cacti (Arroyo-Quiroz, 2010).

Most studies about wildlife trafficking in Mexico involve connections with the United States, which is likely because an overwhelming proportion of illegal wildlife re-exports from Mexico are imported into the United States. (Arroyo-Quiroz, 2010). Although there is evidence of wildlife trafficking between Mexico and some European countries (Arroyo-Quiroz & Wyatt, 2019a; 2019b), these connections have not been explored extensively. Also lacking is consideration of the role of organized crime in Mexican-related wildlife trafficking; organized crime is a general concern throughout Mexico (e.g., Arroyo-Quiroz & Wyatt, 2018; Barth, 2017; Witbooi et al., 2020).

The literature that addresses the relationship between social and cultural constructions of gender identities and the way in which they play out in wildlife trafficking is lacking. Green criminology scholarship can help to advance understanding about crime and harms against wildlife, which include wildlife trafficking (e.g., Sollund, 2019; Wyatt, 2013). Interestingly, although green criminology evolved out of ecofeminist thinking, the theoretical line of scholarship has not been explored to the extent that it might (Sollund, 2019).

DOI: 10.4324/9781003121831-5

This chapter offers a theoretical discussion on wildlife trafficking in Mexico using a feminist and green social justice perspective.

Integrating feminist and green social justice perspectives into discourse about wildlife trafficking helps to elevate diverse types of knowledge generated by diverse voices. Explicitly considering women's experiences in the wildlife trafficking space is important; to ignore it is to impoverish science and its contributions to society. Feminist perspectives remind scholars that science is a political and transformative endeavor, that research is a process—bottom-up, top-down, and in between—and there is a fine line between science and activism within which scientists must situate themselves. In accordance with feminist, green, and social justice, this chapter is intentionally critical, reflexive, and collaborative in its attempt to investigate linkages between gender identities and wildlife trafficking.

Feminist Theory, Political Ecology, and the Usefulness of Gender as a Historical Relational Category

Feminist theory seeks to understand the nature, expression, and transformation of gender inequality. Gender relates to the situated and historical social construction of sexual differences or the ways in which human beings are categorized and socialized according to the social organization of sexual difference within socio-cultural groups (Rubin, 1975). The characteristics linked to sexual differentiation are part of societies' construction of masculinities and femininities, among other gender identities (Serrano Oswald, 2013). Certain characteristics gain prominence and become hegemonic through everyday forms of knowledge and practice, which relate men with masculinity and women with femininity, naturalizing norms over time, and reinforcing systems of power, rights, association, kinship, exchanges, and culture (Scott, 1996). The gendered condition of men and women leads to relational power differences (Segato, 2014). This does not mean that all men and women are in the same situation because some women may have more power than some men (De Luca et al., 2020). Most commonly, societies have a patriarchal social organization, where economic, political, and socio-cultural power lies with men (Serrano Oswald, 2018). Some constructions of masculinity and femininity are so prevalent that human animals assume they apply equally and automatically to the male and female species of the animal kingdom (Ortner, 1974). Biological characteristics of animals are often used to reinforce traits associated with humans, making it difficult to understand that the human condition is historical in nature and has been situated within society and culture (Belausteguigoitia, 2011).

Patriarchy is the historical social system of social inequality, violence, subordination, and injustice that organizes people, power, social spaces, activities, resources, and relations according to political hierarchies in favor of the male gender. It includes processes of alienation, destruction, degradation, oppression, injustice, violence, and the symbolic distribution of space and power (Lagarde, 2015). In the symbolic distribution of space and power in the patriarchy, masculinity represents *homo sapiens* as the ultimate chain of evolution, the master animal closer to reason. Women are considered the *homo domesticus* and closer to procreation, motherhood, and reproduction. The near universal subordination of women to men across cultures since before the agricultural revolution may well be explained by a conception of women as closer to nature and reproduction than men who have led culture and production (Ortner, 1974). According to Pateman (1988), patriarchy represents the deepest ideology of domination in the West, where thousands of years of social division of work have subjected women through social, cultural, and legal sanctions.

The concept of gender derived from feminist theory is a wide umbrella of positions, actions, and critical thoughts seeking liberty and well-being for all human beings. It establishes an intersectional dialogue with other forms of human oppression, and it also understands that human well-being and liberty are related to the well-being of other non-human living beings and the planet. Exploring the conceptualization of masculinity and femininity results in the potential to destabilize gender (Butler, 2006).

Seen from a political ecology perspective, sex–gender systems imprint structural mechanisms for the commodification, circulation, and trafficking of non-human species and particular ways to signify human–natural relations. Following the transversal, systemic, and relational understanding of gender, political ecology has developed into a useful analytical lens that is rooted in inequality, which embodies all forms of inequality with an intersectional understanding (Scott, 1996). Intersectionality means looking at other socially constructed dimensions of inequality and oppression such as race, social class, sexual orientation, age, and education that go hand in hand with gender discrimination, which reinforces it. It is requisite to overcome discrimination, oppression, and inequality.

Characterizing the Context of Engendered Social Justice and Green Justice in Mexico

The 3Rs (i.e., representation, recognition, and redistribution) tripartite model of engendered social justice aims to help assess participatory parity (Fraser, 2000, 2007, 2009). This is highly compatible with the aim of social justice, which is to achieve a just and equitable society where all share in the prosperity of that society that is pursued by individuals and groups through collaborative social action. *Representation* implies self-determination or to affirm as political subject with shared power, civil rights, and political rights. *Recognition* implies considering, recognizing, and affirming an individual or collective subject as a socio-cultural agent beyond essentialisms. *Redistribution* implies restructuring the political economy, overcoming exploitation by social class, and equal sharing of social burdens and benefits. If we consider the three axes of economic redistribution, political representation, and socio-cultural recognition, then the challenges in the Mexican case study are structural, complex, deeply engrained, and even contradictory. Mexico has taken great legal steps toward gender equality. For example, the Law of Public Instruction in 1868 enabled women to access education, to access paid labor in article 123 of the 1917 post-revolutionary Constitution, to obtain education at different levels and in different careers following the principles of the UN Charter, and to enjoy female political emancipation in 1953. Over recent decades, even if women accessed the public sphere massively, which led to significant personal empowerment and transformations in the dynamics of households and labor markets, the overall socioeconomic context has been one of recurrent crises and precarisation, with important gender gaps. A main challenge that has been identified by both UN Women reports on gender representation is a need to incorporate women into public office and to the administration of justice. This argument is critical to this study that involves the administration of green justice that is focused on wildlife trafficking.

To explore these concepts in detail, we obtained profiles of environmental legislative reform promoters in Mexico's Union Congress and Senators and Deputies. The Legislative Information System, Chambers of Deputies, and Chambers of Senators contained information about the work of the environmental commissions between 1997 and 2018. To identify the profile of key actors who have worked in the Ministry of Environment and Natural Resources (SEMARNAT) and the Attorney General for Environmental Protection (PROFEPA), we also accessed official websites and organizational charts regarding wildlife. The names of identified individuals were registered through the National Institute of Transparency, Access to Information and Protection of Personal Data (INAI, 2020). To assess the gender of PROFEPA's field inspectors, we filed an official request for information with the Federal Law of Transparency and Access to Information Public. We also carried out five semi-structured interviews to former and current PROFEPA wildlife inspectors in July and October 2020.

Spheres of Representation and Institutions

Three key aspects make Fraser's feminist social justice perspective (Fraser, 2000, 2007, 2009) useful for analyzing wildlife trafficking in Mexico if one can expand and see it from a political ecology and green justice perspective. First, the fact that unlike other social justice propositions that focus mainly on redistribution, Fraser's proposal incorporates systems of social structures, institutions, and symbolism in an integrated manner, in which each dimension co-creates the other, and change entails systemic transformation. Therefore, despite the aforementioned advances in gender representation, important challenges remain unless all dimensions are addressed. Second, the present late modern epoch with its third wave of globalization is characterized by capital mobility, minority private interests, and State weakness. If the conception of green justice is only centered on multilateral and State institutional frameworks, wildlife trafficking will prevail for a long time unless we address the spheres of redistribution and recognition to feminist and green frameworks openly. Third and last, feminist social justice focuses on the importance of relational fair treatment, or systemic well-being, because over the decades it has become clear that if the focus is not relational and only women or only men have well-being, society as a whole will not progress and equality will not be achieved. Being fair and eradicating violence is a critical invitation to look at ourselves and to understand how patriarchal forms based on violence and oppression cut across the way we treat each other (intra-species), which in turn extends to our inter-species relations. Humans have become increasingly aware of our systemic interconnection and dependence with other non-human life forms and the entire planet. There is also a moral imperative behind fair treatment and respect for all humans that can be extended easily to all other forms of life, which justifies the extension of social justice to green justice.

Governing Structures of Wildlife Management and Use

In Mexico, the exploitation, management, and use of wildlife is carried out through concessions granted by the Federal Executive, according to the rules and conditions established by laws and regulations (Article 27 of the Political Constitution of the United Mexican States [CPEUM]). Key federal laws, regulations, and international wildlife treaties are Supreme Law (Article 133 of CPEUM): the General Law of Ecological Balance and Environmental Protection (LGEEPA), the General Wildlife Law (LGVS) and its regulations, and the NOM-059-SEMARNAT-2010, which lists Mexico's native species of wild flora and fauna and their risk categories and specifications for their inclusion, exclusion, or change of category. The Mexican Federal Legislative Power is in charge of issuing laws and is made up of a General Congress divided into two Chambers: Deputies and Senators (Article 50 of the CPEUM). The Chambers are made up of commissions, which are a form of internal organization of legislative work made up of representatives who have considered specialized arrangements. In Mexico, both Chambers have a commission responsible for ruling, approving, reforming, or repealing wildlife laws.

To comply with national and international environmental legislation regarding wildlife, Mexico has governmental agencies such as the Ministry of Environment and Natural Resources (SEMARNAT) and the Attorney General for Environmental Protection (PROFEPA). SEMARNAT is the main government agency in charge of establishing an environmental protection policy. The General Directorate of Wildlife (DGVS) is responsible for the administration of permits, records, and other documentation that concern the trade, handling, and possession of all wildlife in Mexico. Since 1992, PROFEPA is the administrative body in charge of enforcing compliance of federal environmental legal provisions, and it is the correctional body for those who violate wildlife laws. PROFEPA has the following public officers and administrative units: (1) Attorney; (2) Deputy Attorney General for

Natural Resources; (3) General Directorate of Environmental Impact and Federal Maritime Terrestrial Zone; (4) Directorate General for Inspection and Surveillance of Wildlife, Marine Resources, and Ecosystems; (5) General Directorate of Environmental Inspection in Ports, Airports, and Borders; (6) Legal Assistant Attorney; (7) General Directorate of Federal Crimes against the Environment and Litigation; (8) General Directorate of Environmental Complaints and Social Participation; and (9) Delegations in the federal entities and the Delegation of the Metropolitan Area of the Valley of Mexico (Article 46, Internal Regulations of SEMARNAT).

PROFEPA delegations are responsible for enforcing wildlife legislation. The basic organic structure of each delegation is to have a delegate in charge, who reports directly to the Federal Attorney and will be assisted by subdelegates, deputy directors, heads of department, inspectors, and other personnel necessary for the performance of their duties (Article 68, Internal Regulations of SEMARNAT). In mid-2020, the positions of delegates in most of the states were replaced by the federal government and were called *encargados* (managers). In each delegation, there are: (1) Subdelegation of Inspection of Natural Resources, which include inspectors who supervise wildlife law compliance; (2) Legal Subdelegation, which is responsible for monitoring legal aspects and establishing sanctions on wildlife issues; and (3) Head of Complaints and Communication Department, where complaints made by society on wildlife issues are received and reviewed, which later become possible inspections. The inspectors of the Natural Resources Subdelegation are empowered to carry out surveillance in accordance with applicable legal provisions (Article 47, Internal Regulations of SEMARNAT). They monitor compliance with the provisions of national and international laws and regulations and are empowered to detain possible offenders who violate the provisions on the use of wildlife.

Promoters of Wildlife Regulation Initiatives

Regarding the gender of those who have presented and have approved wildlife regulation initiatives in Mexico between 1997 and 2018, we identified a total of 130 proponents of which 94 were not repeating members. The majority of proponents, who submitted wildlife initiatives (CPEUM, LGEEPA, and LGVS) that were authorized, were men. Only in the case of CPEUM articles related to environmental issues (articles 4, 25, 27, and 73) was it possible to observe an equal number of men and in terms of the number of promoters. It was in the LGVS where the highest number of promoting men were observed (68.8%).

Commissions that Authorized or Rejected Wildlife Regulatory Initiatives

Between 1997 and 2021, a total of eight environmental commissions were identified in the Chamber of Senators and nine in the Chamber of Deputies.

In total, 361 members were identified: 112 in the environmental commissions of the Chamber of Senators and 249 in the Chamber of Deputies. Except for the last legislature (i.e., LXIV, 2018–2021), environmental commissions of the Senate were composed mainly of men, especially at the LIX legislature with the highest number of male members (83.3%) (Figure 4.1). Except for legislatures LXII (2012–2015) and LXIV (2018–2021), these environmental commissions were composed mainly of men (Figure 4.2). In fact, in the

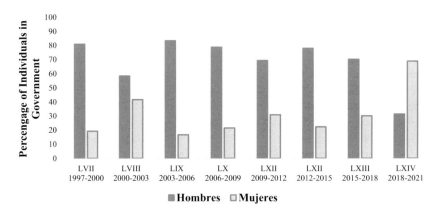

Figure 4.1 Gender representation across environmental commissions according to Legislature, Chamber of Senators, 1997–2021.
Source: Authors' own elaboration with information from SIL (2020) and Cámara de Senadores (2020).

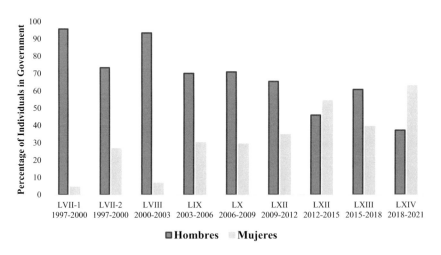

Figure 4.2 Gender representation across environmental commissions according to Legislature, Chamber of Deputies, 1997–2021.
Source: Authors' own elaboration with information from conversations with Instituo Linguistico de Verano (2020) and Cámara de Diputados (2020).

LVII (1997–2000) and LVIII (2000–2003) legislatures, the number of male members were 95.7% and 93.3%, respectively. Between the LVII and LXIV legislatures, there was a decreasing trend in the number of men present in the environmental commissions, and in the LXIV legislature, the number of female members exceeded the number of men. Nonetheless, the majority of those responsible for accepting or rejecting wildlife regulatory initiatives have been men.

Key Positions in the Wildlife Departments of SEMARNAT

We identified a total of 68 individuals who were in key positions of SEMARNAT between 2018 and 2020; there was a greater number of men who occupied key positions (58.8%) than women (41.2%) within the Directorate. There was a larger number of women only in coordination positions, and in leading positions, the number of men and women was the same. Most of the key positions in the Directorate were held by men (e.g., deputy director, boss, liaison); in the sub-directorates, men held the highest number of positions (87.5%). Men held 51.4% of the positions in liaison, where there was the greatest number of individuals.

Key Positions in the Wildlife Departments of PROFEPA

Among positions identified in PROFEPA's headquarters, a total of 58 individuals were identified between 2018 and 2020, of which 65.5% were men. When analyzing the number of men and women in the two main PROFEPA Sub-Attorney Offices (Natural Resources and Legal) and from which wildlife Directorates were derived, most of the key positions were held by men (Table 4.1).

When analyzing the Legal Deputy Attorney General's Office, in particular the General Directorate of Federal Crimes Against the Environment and Litigation, 72.2% of key positions were held by men and 16.7% were held by women; the gender of the remaining 11.1% was not identified. The General Directorate of Environmental Complaints and Social Participation was composed of 55.6% men and 22.2% women. The remaining 22.2% was unidentifiable.

Table 4.1 Gender representation in the two main Deputy Attorney General Offices, PROFEPA headquarters, 2018–2020

Deputy Attorney General Office	Men (%)	Women (%)	Unidentified (%)
Natural Resources	64.5	35.5	0
Legal	66.7	18.5	14.4

Source: Authors' own elaboration with information from INAI (2020).

Table 4.2 Gender representation by key position in Mexico's Deputy Attorney General Offices at PROFEPA headquarters, 2018–2020

Position	Gender (%)		Total (%)	
	Men	Women	Men	Women
Attorney General	0	100	0	3.2
Deputy Attorney General	0	100	0	3.2
Director	80.0	20.0	38.7	9.7
Head of Department	66.7	33.3	12.9	6.5
Coordinator	50.0	50.0	12.9	12.9

Source: Authors' own elaboration with information from INAI (2020).

Table 4.3 Gender representation by key position within Mexico's PROFEPA delegations, 2018–2020

	States (n)	States (%)	Delegate (%)	Subdelegate Natural Resources (%)	Head of Complaints and Communication Department (%)	Subdelegate Legal (%)
Men	19	61.3	74.2	67.7	38.7	48.4
Women	6	19.4	25.8	32.3	38.7	45.2
Men/women	6	19.4	–	–	–	–
Unidentified			0	0	22.6	6.5

Source: Authors' own elaboration with information from INAI (2020).

In the case of the Natural Resources Office, the following wildlife Directorates were identified: (1) Inspection of Marine Protected Areas; (2) Wildlife Inspection; (3) Inspection and Surveillance of Wildlife, Marine Resources, and Coastal Ecosystems; (4) Inspection and Surveillance of Wildlife in Ports, Airports, and Borders; and (5) Environmental Inspection in Ports, Airports, and Borders. The majority of director positions were held by men.

As of 2020, women held the positions of Attorney General and Deputy Attorney General's Offices for Natural Resources. These were the two most relevant positions related to wildlife issues within the Attorney General's Office that were held by women. However, in the directorates and chiefdoms, key positions were mostly held by men (80% in the directorates and 66.7% in the chiefdoms). The coordination positions were held equally by men and women. In relation to key positions held in the delegations of the federal entities in charge of wildlife (Delegate and Deputy Prosecutors) between 2018 and 2020, there was greater representation of men than women; in 61.3% of states, men held the main key positions. Men dominated representation of those responsible for the 32 PROFEPA Delegations between 2018 and 2020. The Delegate position had the largest number of men (74.2%), followed by the position of Subdelegate of Natural Resources, where 67.7% were men (Table 4.3).

In geographical terms, when considering the gender that held the highest number of key wildlife positions in PROFEPA between 2018 and 2020, there was a majority of men in 19 states, a majority of women in six states, and six states held the same number of men and women. In most of the entities that bordered oceans or other countries, key positions were held by men (11 out of 20 entities). Men held key positions in most entities, but this was particularly true in the entities that were key to wildlife trafficking because they were on provincial and international border points. The Delegations where most of the key positions were held by women were the federal entities of Colima, Durango, Jalisco, Sonora, Tabasco, and Zacatecas.

Experiences and Perceptions of PROFEPA's Wildlife Inspectors

All interviewed wildlife inspectors performed the role for approximately five years. Some inspectors considered that if the number of men was higher in key positions, there would be less sensitivity toward wildlife because they perceived women to be more caring and considerate. During the time they worked as wildlife inspectors, they pointed out that the majority of inspectors had been men, which coincided with results from the organizational review. They felt that the treatment of male colleagues and women inspectors was different because there was more consideration given to women. Some interviewees attributed this to the fact that due to the diversity of risk factors in certain geographic areas, it was sometimes disadvantageous to send women into the field.

Interviewees stated that women inspectors were more confident than men when carrying out inspections. They perceived that female inspectors were more dedicated, "taking care of work more" than their male counterparts. They pointed out that more men responded to the calls for new personnel when they were advertised. However, they noted that PROFEPA hired more women, which the interviewees believed was to meet gender quotas. They speculated that fewer women might apply for vacancies or that they might not last long as inspectors because wildlife inspections were conducted in the field under difficult or adverse conditions that were associated culturally with men; women were associated with jobs in the office. Because of this, interviewees noted, men adapted quickly to the work environment and were accepted rapidly by other male inspector colleagues. The interviewees articulated that it was more difficult for female inspectors to adapt to the work environment due to the danger of dealing with offenders or adjusting to late or long hours at work given that "you can take hours on an inspection," which was especially difficult for women that were mothers. Wildlife inspectors claimed that most offenders of wildlife crime that they encountered, especially in the field, were men. They noted that male offenders often treated female inspectors differently, for example, by ignoring female inspectors and only addressing male inspectors, which made it more difficult for female inspectors to perform their duties successfully.

Inspectors noted that in their experience it was very rare for women to commit wildlife crime, which was attributed to the type and place of the activities where wildlife crime was committed. One commented that "the wildlife issue is rude in itself" and takes place in rural areas, where *machismo* is very present and women are assigned home care tasks; for example, they pointed out that when wildlife inspections are carried out in premises such as shops or private homes, in many cases, they were attended by women who commented that their husbands were out working. Interviewees mentioned that during inspections it was more likely to find female offenders involved in activities related to the sale of wildlife or as companions to their husbands or romantic partners during the sale and transfer of wildlife products. They perceived that gender influenced women's involvement in trafficking wildlife. One inspector noted that in regard to wildlife trafficking

> women are in static jobs… in wildlife capturing you need to be able to climb trees and set traps…men develop more physical activities…women work in established trade facilities or are placed to take care of the sale point while men are carrying out extracting activities in the field… the man is the one who obtains and transports [specimens] and women are the ones that sell them.

Inspectors also believed that employing women in wildlife trafficking was a strategy used increasingly by offenders. They mentioned that when public forces, such as police officers, had to intervene in wildlife crime-related offenses, offenders knew that the police generally avoided inspecting or arresting women offenders. Inspectors relayed that police officers have commented that this was because they can be accused of abuse or inappropriate behavior toward the offender, even when policewomen were involved in the apprehension. In addition, in cases of trafficking, commerce, and illegal wildlife breeding, offenders often used family members, such as grandparents, mothers, fathers, uncles, or children, to move products. Inspectors believed this strategy was employed because family members or a woman with children provided greater confidence or even pity to improve sales, and it also helped traffickers evade the authorities during random inspections. For example, one inspector noted that sometimes when we arrived at a site they were received by children, women, or elders; these were vulnerable groups that were sometimes "used as a front" in wildlife trafficking. These people limited their responses to inspectors by saying "that they were sent or that they were put there just to take care." This made it difficult for the inspection process to develop properly because these family members were considered vulnerable.

Reflections on Green Justice and Environmental Protection

The political, legal, and institutional frameworks that guide green justice and environmental protection, which include the fight against wildlife trafficking, in Mexico has been a male-dominated realm historically. In spite of this historical predominance of men, an increasing number of women have been represented at different levels of policy. Between 1997 and 2021, there were important changes in gender quotas and inclusion of women in institutional posts. The current LXIV Legislature in the Chamber of Senators has more women for the first time in history, and the fact that in PROFEPA the Attorney General and the Deputy Attorney are women is unprecedented. Resistance to female participation and *machismo* culture remains visible at the municipal level. Despite the increase in female representation, key positions in wildlife are still led predominantly by males. Mexico's 2020–2021 elections have been the first to enforce paridad transversal (i.e., 50–50 gender quotas in all federal states and municipalities), which will bring changes in representation at the municipal level.

Regulations that relate to wildlife trafficking are the responsibility of the federal government through state delegations and central offices. Mexico is a country with established environmental legal frameworks, although *de jure* conventions and *de facto* reality in terms of green and social justice still require personal, family, community, institutional, and social efforts. Changes that advance gender parity in government are likely to bring resistance, and women might adopt masculine styles and play along with the traditional male agenda. However, time and increasing representation will open up the possibility for change in terms of leadership styles, agendas, priorities, lobbying, and corrupt deals. Perhaps, future research could generate indicators of social justice, gender, and green justice.

Having women in key positions where they can make laws, direct environmental protection agencies, and enforce laws does not translate into gender equality automatically. It is important to include political representation as a transversal axis in supporting green and social justice. History, regulations, field work, and culture itself have operated under masculine conceptions of domination and power. Women also play an important role through soft power in the gender power system that is predatory of all forms of non-human life, which is transversal to all (Lagarde, 2015). Without romanticizing women and considering their intra-generic diversity, womanism will not lead to the depatriarchalization of society automatically, although it is a crucial step forward. Focusing on female representation in positions at all levels is a historical advance because a democracy with transversal parity is a necessary first step in favor of green justice.

It is important not to demonize individual men and women; rather, scholars can consider the anthropocentric–capitalist social representations of nature and property based on patriarchal culture; sex–gender political economy and ecology evolve through identities and cultural processes. Effective and sustainable solutions to wildlife trafficking in Mexico are likely to be collective because its problems are collective. Societal change involves changing conceptions and the practice of a feminist and green social justice, which incorporates the socio-cultural, economic, and political dimensions of recognition, representation, and redistribution.

From a green and gender social justice perspective, it is important to recognize the way in which patriarchal sex–gender systems generate forms of oppression that affect women, men, non-human animals, and the planet in the era of the patriarchal Anthropocene. This suggests that an understanding of mainstream "man"kind and "his"tory that follow an androcentric outlook is important. It is prudent to both widen the lens and to include alternative critical perspectives, dialogues, and narratives and to narrow the focus to look specifically at the ways in which patriarchal anthropocentric civilization models have generated multiple forms of harm. This does not mean that we should consider a women-only or a human-only perspective, and it does not encompass a totalitarian and static perspective that puts all human beings, all women, and all men in the same group. Instead, it enables processes of deconstruction and transformation of oppressive social representations and the construction of alternative ways of understanding identities and cultures. From a green and gender social justice perspective, it means looking at hegemonic sex–gender and anthropocentric practices and considering the different times and spheres of politics: the political realm, policy, polity, politicking, and politicization (Palonen, 2003). Linked to Fraser's conception of social justice (2000, 2007, 2009), it translates into the three interrelated areas of political representation, economic redistribution, and socio-cultural recognition with a transversal green and environmental justice agenda. It includes the formal political realm (i.e., government institutions, legislation, political parties, and public policies) and the more informal organization of political life and participation.

Although it is an important advance, it is not enough to have gender parity in political representation. Economic redistribution across genders and social strata and socio-cultural recognition to deconstruct patriarchal culture and sex–gender systems are also indispensable. Altogether, they create a framework that envisages a more just society that is better able to commit itself transversally to values of sustainability and equality. To address the social, cultural, and economic dimensions of redistribution and recognition, it is important to accept that women participate actively in wildlife trafficking. It will not be possible to overcome wildlife trafficking and transform society without accounting for the roles of women in wildlife trafficking chains.

Literature Cited

Arroyo-Quiroz, I. (2010). *Developing Countries and the Implementation of CITES: A Case Study of Mexico in the International Reptile Skin Trade*. Riga, Latvia: VDM Verlag Dr. Müller.

Arroyo-Quiroz, I., & Wyatt, T. (Eds.) (2018), *Green Crime in Mexico: A Collection of Case Studies* (pp. 149–170). London: Palgrave Macmillan.

Arroyo-Quiroz, I., & Wyatt, T. (2019a). Le commerce et le trafic d'espèces sauvages entre la France et le Mexique. Une étude des asymétries criminogènes. *Déviance et Société, 43*, 569–593.

Arroyo-Quiroz, I., & Wyatt, T. (2019b). Wildlife trafficking between the European Union and Mexico. *International Journal for Crime, Justice and Social Democracy*, *8*, 23–37.

Barth, T.P. (2017). *Organized Crime and the Illicit Wildlife Trade in Mexico*. James A. Baker III Institute for Public Policy at Rice University.

Belausteguigoitia, M. (2011). Hacer y deshacer" el género: Reconceptualización, politización y deconstrucción de la categoría género. *Discurso, teoría y análisis*, *31*, 111–134.

Butler, J. (2006) *Gender Trouble: Feminism and the Subversion of Identity*. London: Routledge.

De Luca, Z.A., Fosado, E.C., & Velázquez, M. (2020) Feminismo socioambiental – Revitalizando el debate desde América Latina. CRIM, UNAM, p. 351.

Fraser, N. (2000). Rethinking recognition. *New Left Review*, *3*, 107–120.

Fraser, N. (2007). Transnationalizing the public sphere: On the legitimacy and efficacy of public opinion in a post-westphalian world. *Theory Culture & Society*, *24*, 7–30.

Fraser, N. (2009). *Scales of Justice: Reimagining Political Space in a Globalizing World*. New York: Columbia University Press.

INAI. (2020). Instituto Nacional de Transparencia, Acceso a la Información y Protección de Datos Personales. Plataforma Nacional de Transparencia. www.plataformadetransparencia.org.mx/web/guest/inicio.

Lagarde, M. (2015) Los Cautiverios de las Mujeres. Madresposas, Monjas, Putas y Locas. Ed. Siglo XXI, p. 624.

Ortner, S.B. (1974). Is female to male as nature is to culture? In Rosaldo, M.Z., & Lamphere, L. (Eds.), *Woman, Culture, and Society* (pp. 68–87). Stanford, CA: Stanford University Press.

Palonen, K. (2003). Four times of politics: Policy, polity, politicking, and politicization. *Alternatives*, *28*, 171–186.

Pateman, C. (1988). *The Sexual Contract*. Londres: Polity Press.

Rubin, G. (1975). El tráfico de las mujeres notas sobre la economía política del sexo. *Nueva Antropología*, *30*, 95–146.

Scott, J. (1996). El género: Una categoría útil para el análisis histórico. In Lamas, M. (Ed.), *El género: la construcción cultural de la diferencia sexual* (pp. 265–302). Mexico City: PUEG.

Segato, R.L. (2014). *Las nuevas formas de la guerra y el cuerpo de las mujeres*. Puebla: Editorial Pez.

Serrano Oswald, S.E. (2013). Migration, woodcarving, and engendered identities in San Martín Tilcajete, Oaxaca, Mexico. In Thanh-Dam, T., Gasper, D., Handmaker, J., & Bergh, S.I. (Eds.), *Migration, Gender and Social Justice* (pp. 173–192). Heidelberg: Springer.

Serrano Oswald, S.E. (2018). Violent social representations and the family as social institution in transition in Mexico. In Brauch, H.G., Spring, U.O., Collins, A., & Serrano Oswald, S.E. (Eds.), *Climate Change, Disasters, Sustainability Transitions and Peace in the Anthropocene* (pp. 153–174). Cham: Springer.

Sollund, R. (2019). *The Crimes of Wildlife Trafficking: Issues of Justice, Legality and Morality*. Routledge: Oxon and New York.

UNODC. (2016). *World Wildlife Crime Report: Tracking in Protected Species* (p. 97). New York: United Nations.

UNODC. (2020). *World Wildlife Crime Report Trafficking in Protected Species 2020*. United Nations. www.unodc.org/documents/data-and analysis/wildlife/2020/World_Wildlife_Report_2020_9July.pdf.

Witbooi, E., Ali, K.D., Santosa, M.A., Hurley, G., Husein, Y., Maharaj, S., Okafor-Yarwood, I., Arroyo-Quiroz, I., & Salas, O. (2020). *Organised Crime in the Fisheries Sector*. Washington, DC: World Resources Institute.

Wyatt, T. (2013). *Wildlife Trafficking: A Deconstruction of the Crime, the Victims, and the Offenders*. Basingstoke: Palgrave Macmillan.

5 Health Implications of Women's Involvement in Wildlife Trafficking in Nigeria

Amelia Ngozi

The Nexus of Human Health and Wildlife Trafficking

Wildlife, or undomesticated animals, can transmit zoonotic diseases to humans through both direct and indirect contact (CDC, 2017; Hilde et al., 2004). Direct contact may include encountering the blood, urine, saliva, mucous, or other bodily fluids of an infected animal through touching or petting, bites, or scratches. Indirect contact may include encountering materials that are contaminated by the infected animal, such as plants, water, or soil (CDC, 2017). Many zoonotic diseases are widely known by the public because they can be highly contagious, fatal, and inflict serious burden on public health officials and medical professionals (Morens et al., 2004). Ebola, severe acute respiratory syndrome (SARS), human immunodeficiency virus (HIV), Lassa fever, and coronavirus disease (COVID-19) are all zoonotic diseases with infections that have resulted in pandemics. Perhaps, in part because zoonotic diseases have the potential to erupt into pandemics and an estimated 72% of emerging zoonotic diseases originate from wildlife (Jones et al., 2008), the intersectionality of zoonotic disease and human–wildlife interactions have been well studied, often from a One Health perspective (e.g., Buttke et al., 2015; Namusisi et al., 2021). The implication of this extant literature is a rich body of evidence about how to better communicate, manage, predict, and assess zoonotic disease risks and risk perceptions (e.g., Decker et al., 2012; Triezenberg et al., 2014).

More recently, scientific inquiry has begun to consider the nature and structure of connections among and between zoonotic diseases and wildlife trafficking (e.g., Aguirre et al., 2021). Wildlife trafficking is a transnational environmental crime that poses diverse harms to indigenous people and local communities, biodiversity conservation, and sustainable development, and it is associated with violence against people and animals that can be associated with other forms of criminality such as drug trafficking (e.g., Gore et al., 2019). Combined, these bodies of knowledge enhance the ability of the public health and medical sectors to manage effectively and efficiently the physical health of humans and wildlife—sometimes called "interspecies health" (Hinchliffe, 2015).

DOI: 10.4324/9781003121831-6

Interestingly, the social and mental health dimensions of zoonotic diseases and wildlife trafficking have not been well studied: a search on Google Scholar (July 2021) yielded no results. This gap in knowledge is noteworthy because social and mental health are correlated closely with physical health (e.g., Ohrnberger et al., 2017), which varies greatly among different groups of people who experience harm differentially, such as women (e.g., Dillon et al., 2013). If positive health outcomes are a policy objective (e.g., Sustainable Development Goal 3: Good Health and Wellbeing), then social and mental health dimensions of wildlife trafficking and zoonotic diseases should be considered alongside physical health. This chapter explores these issues by considering the health implications of women's involvement in wildlife trafficking, using multiple measures of human health and different roles for women in wildlife trafficking.

Assessing Health from a Physical, Social, and Mental Wellness Perspective

Physical health is a concept and state of being that is measured widely on a spectrum of positive to negative. Physical health is tracked at both aggregate and disaggregated levels and is linked to a very large number of independent variables. Good health practices are associated with positive health and often presented within the context of death and disease. Role burdens and physical health have long been explored between women and men (e.g., Verbrugge, 1986).

Social health is often framed around relationships and the ability to adapt to different social situations. There are also social determinants of health, such as economic and social conditions, that influence individual and group differences in physical health status. Women's social roles have been linked to their health status (e.g., Nathanson, 1980) and to indicators such as education level, health care access, and economic stability. Social health has a dynamic relationship with physical health, and many professionals have explored how poor social health resulted in a variety of health issues (e.g., cardiovascular disease, suppressed immune system) (Page, 2020).

Mental health is sometimes defined to include emotional, psychological, and social well-being and ranges from positive to negative. Mental health disorders can be treated in different ways, and symptoms, treatments, and illnesses can all be experienced differentially by women. Some scholars have explored these differences and found that masculinity and femininity affected major risk factors for mental health problems (Rosenfield & Mouzon, 2013).

Women's Roles in Wildlife Trafficking

Agu and Gore (2020) identified six women's roles in wildlife trafficking as an offender, defender, influencer, observer, the person(s) harmed, and beneficiary. These roles were not mutually exclusive or rigid, but rather they were roles that helped to enable inclusion and diversity of thinking about the causes and consequences of wildlife trafficking. These six roles served as the basis for the key informant interviews discussed below: (1) offender, or the primary rulebreaker who engaged in criminal, harmful, or deviant behavior; (2) defender, with the information and formal authority to protect people and wildlife; (3) influencer, or relationship asset with authority, knowledge, and position among a particular group of people; (4) observer, or intentional and unintentional eyewitness to the activities preceding, during, or following wildlife trafficking; (5) person harmed, or vulnerable individual victimized in some fashion by the wildlife trafficking supply chain; and (6) beneficiary, or recipient of indirect or direct advantages from wildlife trafficking. Taken together, exploring the intersectionality of how, when, and why physical, social, and mental health dimensions connect to women's roles in wildlife trafficking offers exciting avenues for research and to more effective interventions designed to reduce risk. To this end, this chapter aimed to advance thinking on the health implications of women's roles in wildlife trafficking.

Key Informant Interviews

Key informant interviews are used in both the health and conservation sectors; the qualitative tool for collecting detailed information can be implemented using face-to-face or electronic communications and conducted with diverse individuals with experience or expertise about a particular topic. The populations of interest for this research were people involved in wildlife trafficking and public health. Both males and females were invited to participate as interviewees to capture a range of perspectives, opinions, and underlying issues. All key informant interviews were recorded digitally so that no important insights were missed and to facilitate qualitative analysis. Thirteen interviews were conducted in October 2020 in Enugu State, Nigeria. Eight illegal wildlife traders (six females, two males) and five public health experts (four females, one male) were interviewed by the researcher and two research assistants. The interviews were done in the local language, Igbo. Their opinions, observations, and perceptions were transcribed verbatim into English using the express scribe transcription software. To ensure quality of the data, all transcripts were reviewed by all of the interviewers. The data were analyzed using reflective thematic analysis (Braun & Clarke, 2019), and the findings were used to complement information from existing literature that are presented under physical, social, and mental health implications.

Physical Health Implications of Women's Roles in Wildlife Trafficking

Key informants discussed a variety of ways in which women's involvement with wildlife trafficking resulted in negative physical health implications; open-ended and unstructured questions guided the conversation. Informants discussed how and when women assumed *offender* and *defender* roles in wildlife trafficking and had direct opportunities to interact with wildlife, wildlife habitat, and people involved in the illegal activity. In this regard, offending and defending women were exposed to zoonotic diseases and physical injury: bites, scratches, lacerations, and stings from animals and insects that carried disease; dangerous and isolated terrain that resulted in falls; exposure to extreme weather, dehydration, and altitude; and lack of suitable equipment and supplies. These direct risks had implications for the physical health of families to which women belonged because women consistently assumed leadership roles in the physical care for nuclear and extended families (Nkangu et al., 2017). Family roles overlapped with wildlife trafficking roles, which increased risk of exposure from offenders and defenders to close family members, particularly with zoonotic disease transmission from shared air, surfaces, food, water, and bodily fluids. Public Health Informant F1 noted, "Women being the caregivers in the home can easily spread any such infections to children and the entire family." Infected family members contributed to community spread when they engaged in educational, employment, or religious activities outside the home.

Women who sold wild meat in markets and served as caterers, restaurateurs, or family cooks also encountered direct physical health risks, according to key informant interviews. Although such individuals may *influence* or *benefit* in some fashion from wildlife trafficking, they disproportionately handled and killed live animals compared to men, cooked, and even consumed wild meat (Volpato et al., 2020). To be sure, not all wild meat was associated with lack of food safety standards and risks, however, Public Health Informant F2 noted,

> because wildlife trafficking is an illegal business, women in it will not be able to apply all the necessary precautions to prevent animal to man transmission of infectious diseases. The wildlife may not receive necessary inoculations against certain diseases. Therefore, if women being family caregivers, use some of the wildlife as food for the family and such wildlife which could harbor disease becomes consumed by family members, the health of such family and the community becomes endangered.

Gender is known to interact with causes and consequences of disease to create different health outcomes for males and females (Vlassoff, 2007). Women are more vulnerable to zoonotic diseases than men (Maher, 2013). HIV/AIDS, a zoonotic disease that originated in primates, is one of the infectious diseases that continue to affect and weaken the condition of women and adolescent girls disproportionately in many resource-limited regions (Gerberding, 2004). Globally, in 2018, an estimated 17.8 million women ≥ 15 years old were assessed as living with HIV, which constituted 52% of all adults living with HIV. An estimated 61% of all young people aged 15–24 years who lived with HIV were young women and adolescent girls, and 59% of new HIV infections among young persons aged 15–24 years occurred among adolescent girls and young women (UN Women, 2018).

Public health officials and medical professionals have long known that consistent exposure to environmental stress and high anxiety conditions result in decreased cardio-metabolic health (e.g., Daniel et al., 2008), and this may differ according to gender (Baldock, 2018). In addition, women often experience greater perceptions of risk, fear of crime, and assessed levels of *victimization* than men (e.g., May et al., 2010; Schafer et al., 2006). Thus, crime and cardio-metabolic outcomes are interrelated (Baldock, 2018).

Female *offenders* may differentially experience the effects of fear of being caught by security personnel or being injured by wildlife, which in turn may influence physical health implications, particularly those that predispose them to increased risk of decreased cardio-metabolic health. Male Bushmeat Seller 2 acknowledged, "some wildlife animals are repulsive under certain conditions at the gunshot, and this is very dangerous" because guns may backfire. Physical confrontations between wild meat *buyers* at markets result in injuries and anxiety about injuries in the process of securing the highest quality or desired products. Female Bushmeat Seller 3 recognized,

> the only challenge I have buying these wild animals is struggling with men at the market to secure the ones I would like to buy. Hunters are few, so bushmeats are becoming scarce, so, you need to be strong to struggle with the men and you can get wounded.

Severe physical injuries lead to chronic pain and deformity of the affected part of the body and, consequently, this impacts a woman's ability to fulfill family responsibilities effectively and efficiently. Tactics to help women avoid physical confrontation have secondary impacts on the amount of time or money they must invest in achieving their objectives. For example, Male Bushmeat Seller 2 noted, "women have difficulties buying these animals from the hunters in the market because we struggle to reach the hunters. So, what they often do to avoid injury is to bid higher than us or buy from us."

Social Health Implications from Women's Roles in Wildlife Trafficking

Key informants offered mostly wide-ranging feedback about the social health implications of wildlife trafficking. They noted that wildlife trafficking attracted a range of penalties and sanctions that were detrimental to the social life of female (and male) *offenders*. When offenders were known to others in their social network, their association with criminality influenced how people in their community related to them. Women are essentialized as being law-abiding and, when they break rules, it negatively impacted her social relationships with friends and family, specifically through increased feelings of shame, shrunken social life, poor school performance for children, and decreased financial support (De Silva et al., 2009).

Offenders' fear of being apprehended by law enforcement authorities \ involved feelings of dependency on certain individuals. Some women were more likely to self-report fear of crime than men, even though they were less likely to experience criminal *victimization* (May et al., 2010), which may have caused women to engage in constrained behavior that included economic or intimate dependency on male protectors (Chan & Rigakos, 2002). Dependency resulted in new exposure to sexual harassment, sexual abuse, and transactional sex risks. Women who sold wild meat, for example, *benefitted* financially from the employment opportunity; however, some business owners intentionally assigned women as market sellers. Male Bushmeat Seller 1 noted, "I can say women experience in this business is that the male customers harass them sexually." Beyond the physical impacts of sexual harassment and abuse, social health impacts included social stigmatization, discrimination, or ostracization in some cultural settings.

Wives of well-known and successful wildlife traffickers, particularly those involved in poaching, were often not well supported in the home. Female Public Health Expert 3 discussed how some women whose husbands were great hunters were often seduced or kidnapped by wealthy men in the community to get the secret of her husband's power. It is believed that great hunters obtained powers by performing some rituals and hunted with some traditional magic or charm to be a successful hunter. In her words,

> in my community, great hunters use charm and perform certain rituals before going inside the forest to hunt wild animals like lion, elephants, etc. For example, there is a leaf in my community that hunters chew that makes them transform in the form of the wild animal so that he can trace the paths and/or habitat of the targeted wild animal. Some charms and rituals are to make them indestructible by the repulsive and dangerous wildlife, and enemies. Some people envy these great hunters and would seek to know the secret of their success, that is, the specific charm he uses or the kind of ritual he performs. So, they target the hunter's wife and either kidnap her or seduce her to get the information from her. Many hunters end up losing their wives to wealthy people in the community, especially if the hunter does not take good care of the wife. But women do not hunt themselves.

Key informants who were questioned about women being offenders in wildlife trafficking expressed they did not perceive them as engaging in aa serious crime; they also stated that it was rare for offenders to be stigmatized or discriminated against. Feelings of pride were also present. Female Bushmeat Seller 1 said,

> I have done this bushmeat business for over 20 years and I am known for it and very proud of it because it has helped me train my children. People

come to patronize me from within and outside of this town and I wash my hands off the blood from raw meat before attending to my customers.

Feelings of ambivalence also existed, which influenced social ties negatively and contributed to depression. Female Bushmeat Seller 3 noted,

> it is a taboo for a woman to carry weapons like a gun and go to the bush to hunt wildlife in Igbo (one of the major tribes in Nigeria) land. We only buy from the hunters or hunters' wives, who sometimes help their husbands with the sale of the dead animal. So, I don't see any health challenge that will be attributed to it.

Under-Considered Mental Health Implications of Women's Roles in Wildlife Trafficking

Importantly, key informants did not directly discuss the mental health dimensions of women's roles in wildlife trafficking. This "under consideration" is noteworthy: physical and social health problems are known to predispose humans to decreased mental health and vice versa (De Silva et al., 2009). One might expect key informants to note chronic pain and traumatic physical injuries that were incurred during offending, defending, observing, or being victimized as part of the wildlife trafficking process. Or how did harms associated with different roles contribute to depression, anxiety, or post-traumatic disorder? Wildlife trafficking defenders (or individuals in other roles) developed negative coping mechanisms to the violence, fear, or victimization they experienced. The implications for mental health included substance abuse or engaging in other wildlife trafficking roles; empirical research would help uncover answers. Defenders were called on to respond to problems where offenders suffered from mental health problems without specialized training (e.g., a mental health issue is not a crime per se) that resulted in physical altercations. Women who served in roles across the wildlife trafficking supply chain were positioned to respond to zoonotic disease infections, physical injury, and deformities with stress, depression, or self-blame (Pulcu et al., 2013).

Depression is a common mental health response known to affect women more than men (WHO, 2020). Mental health problems may serve as a contributing factor for women's engagement in wildlife trafficking roles, particularly if such problems relate to overcoming adversity in other dimensions of life. How, when, and why this relationship emerges is an empirical and important question. A lack of mental health may emerge from a lack of connection skills in their personal and professional relationships, skills to regulate energy and to maintain vitality, and mental skills to drive productive emotions, reactions, and physiology (International Association for Chiefs of Police, 2021).

Addressing Health Implications for Women in Wildlife Trafficking

A deeper understanding about the various health implications of women's roles in wildlife trafficking may help to improve the efficacy and efficiency of responses to the harm and efforts to prevent it. Interventions may underperform or fail to meet expectations if they fail to consider the health implications of women's roles in wildlife. Poor physical, social, and mental health outcomes may reduce women's willingness to participate in wildlife trafficking interventions or attitudes that support wildlife trafficking-related policy interventions. Poor health outcomes could also result to encourage women's engagement in certain wildlife trafficking-related roles.

In many ways, the lack of scientific attention to the health implications of wildlife trafficking on women is a proxy for the general underreaching of women's health. This means that important conditions that may cause problems for conservation, or support solutions, go undiagnosed and unaddressed. A lack of understanding about the differences among and between men and women involved in wildlife trafficking or women involved in different wildlife trafficking roles means that science is not as valid or reliable as it could be. In a world of limited resources and global environmental change, improved representation and engagement have the potential to improve outcomes for people and the environment.

Literature Cited

Agu, H. U., & Gore, M. L. (2020). Women in wildlife trafficking in Africa: a synthesis of literature. *Global Ecology and Conservation*, 23, e01166. Available at https://www.sciencedirect.com/science/article/pii/S2351989420307071#bib19.

Aguirre, A. A., Gore, M. L., Kammer-Kerwick, M., Curtin, K. M., Heyns, A., Preiser, W., & Shelley, L. I. (2021). Opportunities for transdisciplinary science to mitigate biosecurity risks from the intersectionality of illegal wildlife trade with emerging zoonotic pathogens. *Frontiers in Ecology and Evolution*, 9, 15.

Baldock, K. L., Paquet, C., Howard, N. J., Coffee, N. T., Taylor, A. W., & Daniel, M. (2018). Gender-specific associations between perceived and objective neighbourhood crime and metabolic syndrome. *PLoS One*, 13(7): e0201336. Available at https://www.ncbi.nlm.nih.gov/pmc/articles/PMC6062143/.

Braun, V., & Clarke, V. (2019). Reflecting on reflexive thematic analysis. *Qualitative Research in Sport, Exercise and Health*, 11(4), 589–597. Available at https://doi.org/10.1080/2159676X.2019.1628806.

Buttke, D. E., Decker, D. J., & Wild, M. A. (2015). The role of one health in wildlife conservation: a challenge and opportunity. *Journal of Wildlife Diseases*, 51(1), 1–8.

CDC (2017). One Health. Available at https://www.cdc.gov/onehealth/basics/zoonotic-diseases.html.

Chan, W., & Rigakos, G. S. (2002). Risk, crime, and gender. *British Journal of Criminology*, 42, 743–761. Available at https://doi.org/10.1093/bjc/42.4.743.

Daniel, M., Moore, S., & Kestens, Y. (2008). Framing the biosocial pathways underlying associations between place and cardiometabolic disease. *Health Place*, 14(2), 117–132. Available at https://www.sciencedirect.com/science/article/abs/pii/S1353829207000366?via%3Dihub.

Decker, D. J., Siemer, W. F., Evensen, D. T., Stedman, R. C., McComas, K. A., Wild, M. A.,... & Leong, K. M. (2012). Public perceptions of wildlife-associated disease: risk communication matters. *Human-Wildlife Interactions*, 6(1), 112–122.

De Silva, M., MacLachlan, M., Devane, D., Desmond, D., Gallagher, P., Schnyder, U., Brennan, M., & Patel, V. (2009). Psychosocial interventions for the prevention of disability following traumatic physical injury. Cochrane. Available at https://www.cochrane.org/CD006422/INJ_psychosocial-interventions-for-the-prevention-of-disability-following-traumatic-physical-injury.

Dillon, G., Hussain, R., Loxton, D., & Rahman, S. (2013). Mental and physical health and intimate partner violence against women: a review of the literature. *International Journal of Family Medicine*, 1–15.

Gerberding, J. L. (2004). Women and infectious diseases. *Emerging Infectious Diseases*, 10(11), 1965–1967. Available at https://www.ncbi.nlm.nih.gov/pmc/articles/PMC3329060/.

Gore, M. L., Braszak, P., Brown, J., Cassey, P., Duffy, R., Fisher, J.,... & White, R. (2019). Transnational environmental crime threatens sustainable development. *Nature Sustainability*, 2(9), 784–786.

Hilde, K., Anne-Mette, K., & Kjell, H. (2004). Wildlife as source of zoonotic infections. *Emerging Infectious Diseases*, 10(12), 2067–2072. Available at https://www.ncbi.nlm.nih.gov/pmc/articles/PMC3323390/.

Hinchliffe, S. (2015). More than one world, more than one health: re-configuring interspecies health. *Social Science & Medicine*, 129, 28–35.

International Association of Chiefs of Police. (2021). Law enforcement agency and officer resilience training program. Available at https://www.theiacp.org/projects/law-enforcement-agency-and-officer-resilience-training-program.

Jones, K. E., Patel, N. G., Levy, M. A., Storeygard, A., Balk, D., Gittleman, J. L., & Daszak, P. (2008). Global trends in emerging infectious diseases. *Nature*, 451, 990–993. Available at https://www.nature.com/articles/nature06536.

Maher, B. (2013). Women are more vulnerable to infections. *Nature*. Available at https://www.nature.com/news/women-are-more-vulnerable-to-infections-1.13456.

May, D. C., Rader, N. E., & Goodrum, S. (2010). A gendered assessment of the "Threat of Victimization": examining gender differences in fear of crime, perceived risk, avoidance, and defensive behaviors. *Criminal Justice Review*, 35(2), 159–182. Available at https://journals.sagepub.com/doi/10.1177/0734016809349166.

Morens, D. M., Flokers, G. K., & Fauci, A. S. (2004). The challenge of emerging and re-emerging infectious diseases. *Nature*, 430, 242–249. Available at https://www.nature.com/articles/nature02759.

Namusisi, S., Mahero, M., Travis, D., Pelican, K., Robertson, C., & Mugisha, L. (2021). A descriptive study of zoonotic disease risk at the human-wildlife interface in a biodiversity hot spot in South Western Uganda. *PLoS Neglected Tropical Diseases*, 15(1), e0008633.

Nathanson, C. A. (1980). Social roles and health status among women: the significance of employment. *Social Science & Medicine. Part A: Medical Psychology & Medical Sociology*, 14(6), 463–471.

Nkangu, M. N., Olatunde, O. A., & Yaya, S. (2017). The perspective of gender on the Ebola virus using a risk management and population health framework: a scoping review. *Infectious Diseases of Poverty*, 6(135). Available at https://www.ncbi.nlm.nih.gov/pmc/articles/PMC5635524/.

Ohrnberger, J., Fichera, E., & Sutton, M. (2017). The relationship between physical and mental health: a mediation analysis. *Social Science & Medicine*, 195, 42–49.

Page, S. (2020). What is social health? A pillar of wellness workers need to know. https://info.totalwellnesshealth.com/blog/what-is-social-health.

Pulcu, E., Zahn, R., & Elliot, R. (2013). The role of self-blaming moral emotions in major depression and their impact on socio-economical decision making. Available at https://www.frontiersin.org/articles/10.3389/fpsyg.2013.00310/full.

Rosenfield, S., & Mouzon, D. (2013). Gender and mental health. In *Handbook of the Sociology of Mental Health* In: Aneshensel C.S., Phelan J.C., Bierman A. (eds) Handbook of the Sociology of Mental Health. Handbooks of Sociology and Social Research (pp. 277–296). Springer, Dordrecht. https://doi.org/10.1007/978-94-007-4276-5_14.

Schafer, J. A., Huebner, B. M., & Bynum, T. S. (2006). Fear of crime and criminal victimization: gender-based contrasts. *Journal of Criminal Justice*, 34(3), 285–301. Available at https://www.ncbi.nlm.nih.gov/pmc/articles/PMC6062143/#pone.0201336.ref029.

Triezenberg, H. A., Gore, M. L., Riley, S. J., & Lapinski, M. K. (2014). Perceived risks from disease and management policies: an expansion and testing of a zoonotic disease risk perception model. *Human Dimensions of Wildlife*, 19(2), 123–138.

UN Women. (2018). Facts and figures: HIV and AIDS. Available at https://www.unwomen.org/en/what-we-do/hiv-and-aids/facts-and-figures.

Verbrugge, L. M. (1986). Role burdens and physical health of women and men. *Women & Health*, 11(1), 47–77.

Vlassoff, C. (2007). Gender differences in determinants and consequences of health and illness. *Journal of Health, Population, and Nutrition*, 25(1), 47.

Volpato, G., Fontefrancesco, M. F., Gruppuso, P., Zocchi, D. M., & Pieroni, A. (2020). Baby pangolins on my plate: possible lessons to learn from the COVID-19 pandemic. *Journal of Ethnobiology and Ethnomedicine*, 16(19). Available at https://www.ncbi.nlm.nih.gov/pmc/articles/PMC7171915/.

WHO (2020). Depression. Available at https://www.who.int/news-room/factsheets/detail/depression.

Box 2

Voices from the Field

Defenders

Helen U. Agu and Meredith L. Gore

Formal law enforcement authorities such as police, rangers, or customs officials work across multiple jurisdictions to both respond to wildlife trafficking events and prevent the crime from happening in the first place. There are other types of wildlife trafficking defenders besides law enforcement authorities, however. Informal guardians often function at the community level and use non-technical tools such as relationships with local communities. Criminal justice professionals include magistrates, law clerks, and prosecutors – all who work closely with other types of defenders. Some non-governmental organizations help enforce compliance of protected area rules and collaborate with law enforcement authorities on monitoring and crime analysis. Wildlife trafficking defenders are often embedded in the communities they are responsible for. Defenders can facilitate interactions between different wildlife trafficking stakeholders and function as gatekeepers, representatives, liaisons, coordinators, or information brokers.[1] When we asked experts their perceptions of how, when, where, and why women are motivated to serve as defenders in wildlife trafficking (study design, methods, and analysis discussed in Agu et al.[2,3] some mentioned women's qualifications for the role, as these voices from the field illustrate:

> So, it's always difficult if you're dealing with women as they are very strict. And then women have been known to be magistrates and enforcement officers. They are good at prosecuting…they like details and they are always scrutinizing these cases they want to make sure that they take people who have committed a crime to pay for their crimes. Even in terms of work itself, there is a lot of commitment.
>
> (Male 1.3.1)

> If you go to our protected area, it is a good day if you see women are doing their daily activities …because they inform us if people are there to kill elephants. Women automatically report the activities to the office because women are in the protected area collecting grass and fetching water from there then they are a good source of information. Okay. So they are strongly involved with combating poaching or trafficking.
>
> (Male 1.3.5)

Other experts noted different types of inclusion for women in many defender roles, for example:

> So, we now have women as rangers and as community scouts. We have women as wildlife wardens. We have women as prosecutors. We have women as human rights activists too.
>
> (Female 1.3.5)

> We need to get more women into the system [of] protection…and creating awareness in building mentality or attitude, a good attitude to natural resources. So, we must include their involvement in percentage. And we have to equip them with fully fledged logistics to protect them as well. Because they have to be protected before they protect the results if they're exposed, being magnetically or weekly terrain, whatever. The danger is very high.
>
> (Male 12.9.1)

> You know how illegal wildlife trade is affecting not only the wildlife, the species in the wild, but also, you know, security of the area. So, women in many cases have access to security and data. So, they would need to be peaceful. They have to fight illegal wildlife, because in many cases, many places in Africa the lack of security is directly connected to illegal wildlife activities, you know, fueling the war or the whatever it is. So, women if they don't want any problem or any repeat they should cooperate, stop this legal action.
>
> (Male 1.5.6)

> Even in our office we have women who defend against wildlife trafficking. If you go to a protected area, women's daily work helps inform us. If people kill elephants, women are in the protected area collecting grass and water so they are a good source of information especially on the local activity.
>
> (Male 1.5.1)

> There are some women in communities and I'm not talking about police or anything. They're just women from communities that are beginning to say, wait a second, these are resources. These are our resources. These are our animals and we're not going to let you take them.
>
> (Female 1.3.4)

Voices from the field inspire interesting questions about the role of women as defenders in wildlife trafficking. Wildlife trafficking defenders face numerous obstacles in their work, all of which can affect their motivation, retention, and effectiveness. Addressing these obstacles often involves the brokerage of information, relationships, or resources.[4] What conditions best enable defenders to protect humans and wildlife communities, hold individuals accountable, ensure justice, or maintain mutual trust with communities? How do different types of defenders leverage their unique relationships with communities in support of success, and which types of relationships create challenges that need to be overcome? If having women in defender-type roles is a priority, how can women best be recruited and retained? Answers to these and other questions may help reduce harms associated with wildlife trafficking.

Notes

1 Rizzolo, J.B., Gore, M.L., Long, B., Trung, C.T., Kempinski, J., Rawson, B., Huyen, H.T., and Viollaz, J. (2021). Protected Area Rangers as Cultural Brokers? Implications for Wildlife Crime Prevention in Vietnam. *Frontiers in Conservation Science*, 2, 698731. doi: 10.3389/fcosc.2021.698731
2 Agu, H.U., Andrew, C., and M. L. Gore. (2021). Mapping Terra Incognita: an expert elicitation of women's roles in wildlife trafficking. *Frontiers in Conservation Science*. https://doi.org/10.3389/fcosc.2021.683979
3 The Michigan State University Human Subjects Protection Program approved the methods and analysis for the study (STUDY00003659) as exempt under 45 CFR 46.104(d) 2(ii).
4 Rizzolo, J.B., Gore, M.L., Long, B., Trung, C.T., Kempinski, J., Rawson, B., Huyen, H.T. and Viollaz, J. (2021). Protected Area Rangers as Cultural Brokers? Implications for Wildlife Crime Prevention in Vietnam. *Frontiers in Conservation Science*, 2, 698731. doi: 10.3389/fcosc.2021.698731

6 Women, Wildlife Crime, and Sustainable Livelihoods in Cameroon

Eric D. Nana

The Biodiversity and Bioeconomy of Cameroon

The Republic of Cameroon is a lower-middle-income country of over 25 million people located between the Bight of Biafra to the west and the Central African region. It is bordered by Nigeria, Chad, Central African Republic, Equatorial Guinea, Gabon, and Republic of Congo. Its total area is 475,442 km² and is endowed with a rich biological diversity and primary ecosystems, such as savannah and tropical rainforest (MINADER, 2015). Its rich biodiversity is characterized by a high level of endemism and, as part of the Congo Forest Basin, Cameroon is home to rich cultural diversity as well. It is home to approximately 8,300 species of plants, 335 species of mammals, 542 fresh and saline water fish species, 950 bird species, 330 species of reptiles, and 200 amphibian species (Delancey & Delancey, 2000; Fishpool & Evans, 2001). Cameroon is one of the most biodiverse countries on the African continent; many of its species are found nowhere else on earth and many of its ecosystems are threatened. Cameroon ranks fifth in Africa for faunal richness and fourth for floral richness (Tchouto et al., 2006; Watson et al., 1995).

Cameroon's rich biodiversity is a source of food, fuel, and livelihood for millions of men and women – a bioeconomy; however, this biodiversity is threatened by unsustainable logging, poaching, and climate change (MINADER, 2015). Cameroon's economy has long depended on its natural resources (e.g., timber, minerals), and its biodiversity is currently affected by natural resource use; this is further exacerbated by few job opportunities outside the agricultural sector. The overall number of poor in Cameroon increased by 12% between 2007 and 2014, and poverty is increasingly concentrated, with 56% of the poor living in the northern regions (World Bank Group, 2019). Cameroonian women are more vulnerable to socioeconomic conditions compared to men (Tedou et al., 2011). Women face high levels of discrimination and gender inequality in social institutions (Djapou & Chimene, 2017). Women's participation in local government (8% of local councilors, 2% of political party leaders) is lower than that in national government (31% in Cameroon's National Assembly). Gender inequality in leadership positions, lack of women's economic empowerment, and the ongoing

DOI: 10.4324/9781003121831-8

civil crisis in the English-speaking part of the country has contributed to closure of schools, violence and armed conflicts, incarceration of civilians, and displacement of thousands of people who have fled their homes (UN Women, 2021). These have certainly contributed to women's engagement in illicit commercialization of wildlife through poaching as a way to cover intermittent needs.

Threats to Biodiversity from Conservation Crimes

A subset of Cameroonian people derive their livelihoods from forests, especially in rural areas. Forest products, such as timber and non-timber forest products, constitute a valued source of income. Bushmeat plays a very significant role in the health and wellbeing of local people and their families (Cawthorn & Hoffman, 2015; Ichikawa et al., 2016; van Vliet et al., 2017) despite the fact that many of these species (e.g., white-bellied pangolin [*Manis tricuspis*], Gabon viper [*Bitis gabonica*], and rock python [*Python sebae*]) are classified as threatened on the red list of species of the International Union for the Conservation of Nature (IUCN) and are also protected by national law. Killing, transporting, and consuming these species are illegal. Bushmeat trafficking is highly valuable, and some estimates value the trade at between USD $5–20 billion per year worldwide (Milliken, 2014). In the Central African region, prices vary depending on the species (or animal part), but can reach values up to a final market price of USD $400,000 for a single gorilla (Clarke & Babic, 2016). Because of the very lucrative nature of this illicit business, gender inequalities that inhibit women from obtaining jobs, and the chronic economic crisis that has crippled Cameroon's economy since the late 80s, it is understandable that some women have entered into wildlife crime as a means to achieve a sustainable livelihood.

Three Channels of Wildlife Crime in Cameroon

Currently, wildlife crime in Cameroon can be distinguished based on three different channels (Table 6.1): rural consumption, urban consumption, and foreign consumption. There are three main actors: harvesters/poachers (for subsistence or commercial purposes), retail buyers, and end consumers (Lescuyer & Nasi, 2016). Regarding the channels, rural consumption is mostly for people in money-poor communities who live in rural areas and who cannot afford to buy high-value bushmeat (i.e., wildlife species that have a good market appeal and are sold at much higher prices), such as brush-tailed porcupine (*Atherurus africanus*), white-bellied pangolin (*Manis tricuspis*), warthog (*Phacochoerus aethiopicus*), duikers (*Cephalophus dorsalis*), Gabon viper (*Bitis gabonica*), rock python (*Python sebae*), Old World monkeys (*Cercopithecus* spp.), and elephant (*Loxodonta africana*). In the case where a project (e.g., government construction) is carried out in a rural area and workers are brought in from different localities, or even countries, these high-value bushmeat species

Table 6.1 Wildlife trafficking channels that show actors in different categories observed in Cameroon.

Category	Wildlife Trafficking Channels				
	Rural Consumption		Urban Consumption		Foreign Consumption
Actors	Poachers[a]	Retail buyers[a,b]	Retail buyers[b]	End consumers[a,b]	End consumers[a,b]
Money-poor citizens	X	X			
Project workers (nationals and/or expatriates)		X	X	X	
Middle class & rich citizens			X	X	
White-collar poachers					X

a Men.
b Women.

are sold to the workers. As actors in this channel, men are the sole poachers, but both men and women are retail buyers.

Urban consumption is usually for city dwellers who belong to the middle class, upper middle class, and the rich. These are mostly the end consumers, and their love for the taste of bushmeat keeps the trade alive. Women make up the vast majority of retail buyers in the urban consumption category. Foreign consumption is usually for animal parts (e.g., pangolin scales, elephant tusks), which are destined for Southeast Asian countries like China. The amount of money involved is such that highly trained poachers who operate with military automatic machine guns have entered this trade. In 2012, for example, poachers on horseback (reportedly Sudanese horse militias) killed several hundred elephants in northern Cameroon within a few months (Mongabay, 2012) and escaped with the ivory that was possibly destined for international markets. Beyond this three-channel conceptualization, little is known about the organizational structure of wildlife crime in Cameroon. Regarding actors in the foreign consumption channel, poachers are all observed to be men and are part of international trafficking.

Wildlife crime is often thought of in Cameroon as a crime of hegemonic masculinity, and this has led to little attention directed to the gendered dimension of this crime. One reason may be because the vast majority of offenders that were involved in wildlife trade and in the killing of endangered animals were men. Even when offenders were caught and paraded on television, it is observed that they were all men. However, there were a few cases where women were observed to be offenders, and this was usually at sale points in urban areas.

Many reasons can be advanced to explain why these women at sale points take part in wildlife crime in Cameroon. They are underrepresented in decision-making spheres at all levels, if not restricted completely. They do not have the same legal rights to land, property, and resources as men, and this strips them of critical means for survival and resilience. Although women have crucial roles and responsibilities in natural resource management, their contributions, innovations, and leadership are often overlooked. The result is a lost opportunity for environmental initiatives to achieve multiple benefits and to increase effectiveness, given that gender equality leads to more successful, efficient, and equitable development outcomes. To better understand wildlife crime and the involvement of women, this chapter explores (1) how wildlife crime is organized in Cameroon and (2) the various roles that women play across the supply chain.

Conceptual Framework

Agu & Gore (2020) created a conceptual framework of roles that women play in wildlife trafficking. If we consider the variety of roles that women experience in wildlife trafficking, this may help to enhance crime prevention policies and programs. To be effective, such activities ideally should consider the gendered dimension of the issue and the intersectionality of gender with wildlife crime. Women can be powerful agents of change, and changing the trajectory of wildlife crime in Cameroon is a high policy priority (MINE-PRD, 2020). Six primary roles in wildlife trafficking and the extent to which they could be applied to a woman are:

Offenders – individuals or groups "doing" the criminal, harmful, or deviant acts. For example, retail buyers.
Defenders – individuals or groups with formal or informal authority charged with guarding or protecting people and animals across the wildlife trafficking supply chain. For example, female rangers and women working for non-governmental organizations involved in fighting wildlife crime.
Influencers – individuals or groups linked by various mechanisms of connection to wildlife trafficking and with the capacity to stimulate and to suppress. For example, romantic partners.
Observers – individuals or groups that are eyewitnesses to the activities of, and actors involved in, wildlife trafficking, either intentionally or unintentionally. For example, other women traders.
Victims – individuals or groups made vulnerable by wildlife trafficking. For example, wives and daughters of poachers caught and jailed.
Beneficiaries – individuals or groups that derive indirect or direct benefits from wildlife trafficking, such as families of poachers and retail buyers.

Data Collection

Mixed methods were used to achieve objectives, which included multiple sources of data to help triangulate insights.

- Conservation and development organization and national administration websites were reviewed to understand local involvement of various roles for women involved in wildlife trafficking (e.g., Support Service to Local Development Initiative, Last Great Ape Organization [LAGA], Congo Basin Forest Partnership, and World Wildlife Fund for Nature [WWF]); reports, technical documents, and website content were reviewed.
- Key-informant interviews with high-ranking officials (five men, one woman) in the relevant ministries and departments, traders in open markets, restaurants, and cafeterias (180 total [120 women, 60 men]), traffickers (ten men), rangers (five [one woman, four men]), and experts from academic institutions and non-governmental organizations (ten [two women, eight men]). Due to the sensitive nature of wildlife trafficking, informants were selected by random sampling methods described by Fimbel et al. (2000), the randomized response technique (Blair et al., 2015; Bose, 2015) to administer open-ended questionnaires for a maximum of 30 minutes per person.
- One hundred and fifteen sale points (90 restaurants and cafeterias, 25 open markets) were also visited once during early morning hours (between 5 and 6 am) or midday depending on the type of sale point in the main cities of Yaoundé and Douala with the help of maps of these cities obtained from the National Institute of Cartography. Each point was visited once and covert observations made with geographic coordinates taken using a Global Positioning System. Sale points were classified as: (1) cafeterias (visited during midday hours), where food is sold in a kiosk on the street, but comfort is limited and prices are low, (2) restaurants (visited during midday hours) which are of different standing with durable infrastructure. These are relatively more comfortable compared to cafeterias and prices are higher, which are meant for average to wealthy citizens, and (3) market retailers (visited during early morning hours), who are people who sell bushmeat on selling tables or on the market floor. The number and sex of sellers was noted on paper, and the sellers were queried about the origin and type of their bushmeat suppliers.

Two approaches were used to describe and to quantify the economic importance of wildlife crime to women: a market chain analysis (MCA) to define the roles within the bushmeat trade and a cost–benefit analysis (CBA) as a systematic process to calculate and to compare benefits and costs of the bushmeat trade (Kaplinsky & Morris, 2001). Within CBA, inflows consisted of bushmeat sales and the roles associated with this activity. The economic benefits and costs of bushmeat trade revealed the contribution of bushmeat to livelihoods (i.e., the consumption of bushmeat by both urban and rural populations). The analysis presented here covered only one year and does not predict the evolution of the bushmeat trade over the medium and long term. Data used was obtained from key-informant interviews and secondary sources, that is, scientific publications and student theses on bushmeat extraction and trade in Cameroon. Scientific references for Cameroon were selected from Taylor et al. (2015). PhD and MSc theses were consulted from research institutions and universities in Cameroon.

Describing Wildlife Crime in Cameroon

Trafficking Routes

The different wildlife trafficking channels in Cameroon help illustrate nuanced dimensions of supply and demand chains that are much higher in urban compared with rural areas and that are more lucrative for foreign consumption (e.g., pangolin scales, elephant tusks) (Bahuchet & Ioveva-Baillon, 1998). In urban areas, consumers who have a much higher purchasing power are supplied either by buying at specialized markets or by consumption in restaurants or cafeterias. At the rural level, supply of bushmeat can be done directly by hunters and retail buyers or by display of bushmeat at the roadside for travellers to see and to bargain. It is important to note that wildlife is killed at this rural level in protected areas or their surrounding buffer zones and, as such, the direct supply to consumers is less complicated. Transport routes that supply bushmeat to urban areas are: (1) by train from the northern part of the country, mainly from Ngaoundere in the Adamawa region, and (2) major motorable roads; from the western part of the country (e.g., Bafoussam, Bamenda, Buea), southern region (e.g., Ebolowa, Sangmelima), and eastern region (e.g., Bertoua). These transport routes mostly take bushmeat to the two biggest cities in Cameroon: Yaoundé, the capital city, and Douala, the economic capital where purchasing power and demand for bushmeat are highest among individuals with disposable income.

Control Points

Traffickers of bushmeat and other forest products that were obtained illegally bribed their way to the various sale points. They paid money, which varied between USD $10 to $100 (depending on the species they are transporting and also on their bargaining power), to law enforcement agents who were at the different fixed control points in rural areas and at the entrance of main cities. This amount paid along the route of transport was added to the final cost of the bushmeat that increased the price of bushmeat in urban areas even more. These law enforcement agents considered the money they received from traffickers as a bonus necessary to cover their family needs, given their low salaries.

Sale Points

In both Yaoundé and Douala, by midday, hundreds of women served approximately 2,000 bushmeat dishes daily in restaurants and cafeterias or sold bushmeat in specialized markets. The prices of the bushmeat or dish depended on the species. Wildlife species that were most frequently found at these sales points can be classified into three categories: (1) those that were totally protected under Cameroonian law and classified as list A species: white-bellied pangolin, elephant meat, and Old World monkeys; (2) those where consumption required an authorization from the relevant government ministry of wildlife and forestry – list B: warthog, bush pig (*Potamochoerus porcus*), duikers, monitor lizards (*Varanus niloticus*, *V. griseus*), Gabon viper, rock python; and (3) those that were only partially protected and where consumption was based on a limited number of individuals killed per month: brush-tailed porcupine, cane rat (*Thryonomys swinderianus*). Those who frequent these sale points belonged to all social classes. One interesting observation was that law enforcement officers and government agents were among the most valued clients in these restaurants and cafeterias.

Economic Importance of Wildlife Crime to Women

The bushmeat trade in the main cities of Cameroon is scarcely controlled by the state and it is regulated by economic operators on the supply side at the rural level, that is, by poachers/harvesters who are all men. At the rural level, there are three options for selling bushmeat: (1) direct sales to rural inhabitants; (2) display of game at the roadside for sale to passing travellers; and (3) sale to a trader/dealer who comes to take his order directly from the hunter. A third of the bushmeat inflows in urban areas in Cameroon does not go through bushmeat markets and this makes it difficult to quantity the amount of bushmeat consumed or to be sure of the bushmeat flows outside market channels (Bahuchet & Ioveva, 1999). Based on the low estimation of the direct flows of bushmeat, the amount of bushmeat sold, mostly by women, in restaurants and cafeterias or in specialized markets in the major cities of Yaoundé and Douala can thus be estimated to about 11,000 tons per year. The prices of the bushmeat vary due to several factors such as the state of the game (fresh or smoked), the species, and the extent of processing but overall, as a generic figure, an average price for bushmeat sold on urban markets without reference to species and the state of the game was USD $6/kg.

The operating costs of the bushmeat value chain depend on the following expenditures: (i) bushmeat bought in rural areas from poachers/harvesters by women retail buyers at an average price of USD $2.7/kg; (ii) the average transportation cost to rural areas around the main cities of Yaoundé and Douala (on the basis of the current prices of public and informal transport in Cameroon) to purchase the bushmeat is estimated at USD $21.7 per trip; (iii) costs for accommodation, food, and communication were estimated at USD $9 per day for a retail buyer and subject to a value-added tax of 19.25%

(Lescuyer & Nasi, 2016; MacDonald et al., 2012); (iv) the bushmeat may be subject to waste either because of the deteriorating condition of the meat before it reaches urban markets, or because of formal seizure by the administration. This increases the cost to the retail buyer but due to a lack of empirical data on such waste, a 10% increase in the total operating cost was assumed (Lescuyer & Nasi, 2016). An annual extrapolated to all bushmeat sold by retail buyers to end consumers in the main cities of Yaoundé and Douala in Cameroon reveals the net financial benefit estimated at USD $15 million per year. This helps explain why wildlife crime is very lucrative, especially to women who make up majority of the retail buyers.

Characteristics of Women Involved in Wildlife Crime

This chapter shows there are three main actors as mentioned above, and these actors played different roles that were gender-based. Women belonged to two groups of actors (retail buyers and end consumers). Following the characterization of roles according to Agu & Gore (2020), the following roles were observed in order of importance (see Figure 6.1):

Beneficiary (40%) – women who played this role were both retail buyers and end consumers and were the most common type. They derived direct or indirect economic benefits from wildlife trafficking. Many women in rural and urban areas were identified as deriving direct economic benefits from wildlife trafficking.

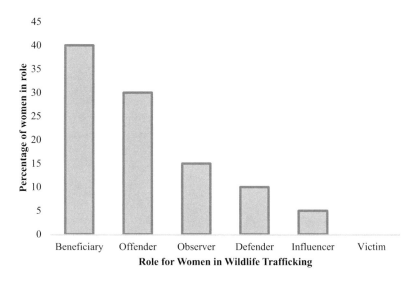

Figure 6.1 Percentages of women observed to play different roles in wildlife crime in both urban and rural areas in Cameroon in order of importance. Most women benefited directly or indirectly from wildlife crime

Offenders (30%) – women who played this role were retail buyers and were second highest in number. They occupied a powerful position within the circuit because they sustained the supply chain with the money they injected into the circuit that served as the main source of revenue for poachers. Some obtained bushmeat from illegal early morning markets (open at 4:30 am and close at 5:30 am) in the eastern region of Cameroon (Yokadouma) or from poachers who supplied them directly. These early morning markets were almost all in rural areas and served as exchange grounds between poachers and retail buyers. These were the places where species that were protected totally by Cameroonian law were transacted unlike the midday markets where only species not protected or partially protected by law were sold.

Observers (15%) – these were women who were eyewitnesses to the activities of traffickers in rural areas and in local bushmeat markets. They did not belong to any of the three channels of actors and did not benefit personally from the illicit traffic. They preferred to be indifferent to it and considered that those involved did it as a means of survival. These were women who sold legal products in a market, such as cloth, spices, or vegetables/fruits both in rural and urban areas.

Defenders (10%) – these were women involved in law enforcement who were known locally as Ecoguards. They were staff of the Ministry of Wildlife and Forestry who were deployed to the different protected areas, and they controlled stations in rural areas around the country. They had the authority to arrest wildlife criminals. Others were women who worked with non-governmental organizations such as LAGA, WWF, and IUCN and were very active on the ground because they obtained information that they relayed to law enforcement agents and, occasionally, took part in arrests. Some of the women who played this role were also end consumers.

Influencers (5%) – These could be very wealthy women with white collar jobs, such as lawyers, medical doctors, wives of wealthy men who lived in the major cities, or women who lived in money-poor communities in villages with local men, who undertook poaching for subsistence and commerce. These women saw poaching as an opportunity to either make more money or get a balanced diet. Some even occupied top administrative positions.

Victims (0%) – No woman was identified who had been made vulnerable or harmed by wildlife trafficking.

Trafficking Routes

The high demand for bushmeat in urban areas like Yaounde and Douala can be explained by the fact that residents in these urban areas had a higher purchasing power while rural inhabitants desperately needed the money to make a living. Van Vliet & Mbazza (2011) found that higher wages in urban areas explained the higher proportion of bushmeat sales (70%) compared with rural areas (30%). The two main cities in Cameroon, Douala and Yaoundé, had 24% of Cameroon's population, but produced a total of 44% of gross domestic product (World Bank, 2016). This implied that these two cities were approximately 1.8 times more productive than the rest of the country with large improvements in the urban poverty rate, which had fallen from 17.9% in 2001 to just 8.9% in 2016 (World Bank, 2016). Bahuchet & Ioveva-Baillon (1998) and Ellis (2000) found that women throughout Cameroon occupied a powerful position within the circuit and made up 84–85% of retail buyers of bushmeat at sales points in major urban areas like Yaoundé. Since the mid-90s, bushmeat is no longer regarded as luxury meat for wealthy people, but was affordable by even the lower middle class, which has caused an increase in the number of consumers by >10% (Edderai & Dame, 2006; Wilkie & Carpenter, 1999). Conservation has long focused on protected areas or biodiversity hotspots, but this research shows again that there is an urban dimension to local conservation. This urban dimension is so important to wildlife conservation and also to the role of women because it drives or sustains the intersectionality of urban and rural conservation.

Roles of Women

Data obtained from this study showed that two of the most common roles women played were beneficiary and offender. As beneficiaries, women may benefit directly from wildlife crime economically as retail buyers in local markets and smugglers in restaurants, or indirectly from a nutritional perspective as end consumers when they prepare a meal of bushmeat at home for their families (Cawthorn & Hoffman, 2015; Coad et al., 2010). Many women who consumed bushmeat in Cameroon, especially threatened bushmeat species such as pangolins and elephants, were not even aware they transgressed the law by eating such species. When asked if they knew anything concerning the national legislation on wildlife, many responded that they did not. There is a need for sensitization of end consumers and traffickers of bushmeat to focus their "take" on the most common species, which are also the most resilient, because such species are usually not threatened, and there would be no need to ban all bushmeat trade (van Vliet & Mbazza, 2011).

As offenders, women in Cameroon followed a general tendency among African women of being mostly buyers and sellers (retail buyers) of bushmeat in local markets (Belo-Osagie, 2017; Wairima, 2016). International export of bushmeat or live animals was assured by men in Cameroon, but there were few cases of women who were involved with their partners as accomplices (Cooney et al., 2017). Their role as buyers and sellers made them the backbone of wildlife traffic locally because they provided poachers with the needed resources and motivation to continue their indiscriminate killing of game, and they also provided end consumers the meals they wanted. All women buyers and sellers of bushmeat who were interviewed accepted that they knew they violated the law, but put the blame on the government who they accused of being responsible by not providing them with alternative economic activities.

Getting the women involved to indulge in alternative livelihoods will require political will and investment at multiple levels: legal, financial, educational, and social. This requires further research to help provide answers. This chapter helps to provide information about entry points for different types of intervention – and also highlights where women entered and exited conservation crime roles. Cameroon recently developed a national policy on gender equality to remedy inequalities between men and women in different sectors (MINEPRD, 2020). The policy document recommends active support for women's businesses, which could have positive implications for wildlife trafficking by raising awareness of women's rights and gender in a more general way. However, it should be noted that the gender policy does not address directly the customary barriers to gender equality or the inconsistencies among land, forestry, and other laws that affect natural resources. Although this is an important step in achieving gender equality, national gender policy could be complemented by gender-based strategies, such as those outlined in the national development strategy document 2020–2030 (MINEPRD, 2020).

As defenders, women served as law enforcement agents who were known locally as Ecoguards. They were staff of the Ministry of Wildlife and Forestry who were deployed to the different protected areas and control stations around the country. These women were few in number and on the frontlines; they risked their lives to tackle poaching and the illegal wildlife trade. The reason why there were so few women who served as rangers (Ecoguards) in Cameroon was because the job has been a male's job traditionally, and it is quite risky. It was not appealing to women because of the remote working conditions and a meager salary. Today, more and more women have become Ecoguards, and a recent study by Strobel (2019) showed that women's inclusion in law enforcement against wildlife crime is a key ingredient that has brought organizational change, better use of decision-making by the force, and a better response to institutional and community needs. Other women who also played the role of defenders worked with non-governmental organizations such as LAGA, WWF, and IUCN. Their role in these organizations was to undertake outreach and awareness-raising campaigns in local communities adjacent to protected areas and to support compliance with rules and norms.

As influencers, these were women who exerted financial or social pressures on individuals involved in wildlife trafficking. Financial pressure was usually exerted by women who occupied top administrative positions or who were wives of top officials who employed poachers to hunt for them. On the other hand, social pressure was exerted by women on men when they used food or the promise of a romantic relationship to drive these men to undertake illegal hunting. Such incentives used by women appeared to be the main drivers of poaching, according to the reports reviewed for this analysis.

Women as observers were eyewitnesses to the activities of traffickers and of local bushmeat markets. They did not benefit personally from the illicit traffic and preferred to be indifferent to it. They usually preferred to have no opinion about poaching. Women usually became victims or were made vulnerable by wildlife trafficking because their husbands or partners were put in jail or died as a result of trafficking (Hübschle & Shearing, 2018; Masse et al., 2018). These women were heavily affected socially and seldom accepted an invitation to take part in any interview. This was probably the main reason that we could not identify any woman during interviews or from studies in the literature about women who were made vulnerable by wildlife trafficking.

Conclusion

Wildlife trafficking is often thought of in Cameroon as a crime of men and that women seldom participate in it because of the risks involved. This chapter has argued that the roles in wildlife trafficking were gender-based, and the very lucrative nature of this ecological crime attracted more and more women because of the high levels of social discrimination against them and few job opportunities available in-country. Many women may choose wildlife trafficking as a means of livelihood and were very active players in it despite the risks. The role women played in this study, for example and although limited to retail buying, enabled them to sustain the traffic in-country, and therefore, women were central to it.

The roles women played in wildlife crime include roles both in favor of it and, at times, also against it. There were women on the frontlines in the fight against wildlife crime as defenders either as rangers or as conservation activists in non-governmental organizations. These women risked their lives to tackle poaching and the illegal wildlife trade, and they were often posted far from their families and loved ones. These women should be encouraged and should receive some kind of motivation, financially or psychologically, because they proved daily that men alone cannot stop wildlife trafficking.

Acknowledgments

I am grateful to Nkemnyi Standly and Ngansop Eric from the Institute of Agricultural Research for Development for their assistance during data collection.

Literature Cited

Agu, H. U. and Gore, M. L. (2020). Women in wildlife trafficking in Africa: a synthesis of literature. *Global Ecology and Conservation*, 23, e01166. doi.org/10.1016/j.gecco. 2020.e01166.

Bahuchet, S. and Ioveva-Baillon, K. (1998). Le role of street restaurants in the city supply-chain of bushmeat: the case of Yaoundé (Cameroon). In D. Bley, J. Champaud, P. Baudot, B. Brun, H. Pagezy and N. Vernazza-Licht (eds.), *Cities of the South and Environment*. Editions du Bergier, Marseille (In French), pp. 171–182.

Bahuchet, S. and Ioveva, K. (1999). De la forêt au marché: le commerce de gibier au sud Cameroun. In S. Bahuchet, D. Bley, H. Pagezy and N. Vernazzalicht (eds.), *l'homme et la forêt tropicale*. Edition du Bergier, Chateauneuf de grasse, pp. 533–558.

Belo-Osagie, M. (2017). Women Must Guide Africa's Conservation and Development | African Wildlife Foundation. Retrieved 13 October, 2020, from https://www.awf.org/ blog/women-must-guide-africas-conservation-and-development.

Blair, G., Imai, K. and Zhou, Y.-Y. (2015). Design and analysis of the randomized response technique. *Journal of the American Statistical Association*, 110, 1304–1319.

Bose, M. (2015). Respondent privacy and estimation efficiency in randomized response surveys for discrete-valued sensitive variables. *Statistical Papers*, 56(4), 1055–1069.

Cawthorn, D. M. and Hoffman, L. C. (2015). The bushmeat and food security nexus: a global account of the contributions, conundrums and ethical collisions. *Food Research International*, 76, 906–925. https://doi.org/10.1016/j.foodres.2015.03.025.

Clarke, A. and A. Babic (2016), "Wildlife trafficking trends in sub-Saharan Africa", in *Illicit Trade: Converging Criminal Networks*, pp. 57-77. OECD Publishing, Paris, https://doi.org/10.1787/9789264251847-6-en.

Coad, L., Abernethy, K., Balmford, A., Manica, A., Airey, L. and Milner-Gulland, E. J. (2010). Distribution and use of income from bushmeat in a rural village, Central Gabon. *Conservation Biology*, 24, 1510–1518. https://doi.org/10.1111/j.1523-1739.2010.01525.x.

Cooney, R., Roe, D., Dublin, H., Phelps, J., Wilkie, D., Keane, A. and Biggs, D. (2017). From poachers to protectors: engaging local communities in solutions to illegal wildlife trade. *Conservation Letters*, 10, 367–374. https://doi.org/10.1111/conl.12294.

Delancey, M. W. and Delancey, M. D. (2000). *Historical Dictionary of the Republic of Cameroon*. Third edition. Lanham, Scarecrow Press, London, p. 359.

Djapou, F. and Chimene, D. (2017). Dealing with Challenges on Gender Inequality in Cameroon. Humanistic Management Association, Research Paper Series No. 17-33. http://dx.doi.org/10.2139/ssrn.2985610.

Edderai, D. and Dame, M. (2006). A census of the commercial bushmeat market in Yaoundé, Cameroon. *Oryx*, 40, 472–475. https://doi.org/10.1017/S0030605306001256.

Ellis, C. M. (2000). An integrated model for conservation: case study on the role of women in the commercial bushmeat trade in Cameroon. M.Sc. thesis, York University, North York, Ontario, p. 186.

Fimbel, C., Curran, B. and Usongo, L. (2000). Enhancing the sustainability of duiker hunting through community participation and controlled access in the Lobeke region of southeastern Cameroon. In J. G. Robinson and E. L. Bennet (eds.), *Hunting for Sustainability in Tropical Forests*. Columbia University Press, New York, pp. 356–374.

Fishpool, L. D. C. and Evans, M. I. (2001). *Important Bird Areas in Africa and Associated Islands: Priority Sites for Conservation*. Pisces Publications and BirdLife International, Newbury and Cambridge (BirdLife Conservation Series No. 11).

Hübschle, A. and Shearing, C. (2018). Ending Wildlife Trafficking: Local Communities as Change Agents. Global Initiative against Transnational Organized Crime. Retrieved 13 October, 2020, from http:// globalinitiative.net/wp-content/uploads/2018/08/TGIATOC-Wildlife-Trafficking-Report-WEB-4.pdf.

Ichikawa M., Hattori S. and Yasuoka, H. (2016). Bushmeat crisis, forestry reforms and contemporary hunting among Central African forest hunters. In V. Reyes-García and A. Pyhälä (ed.), *Hunter-Gatherers in a Changing World*. Springer, New-York, pp. 59–75.

Kaplinsky, R. and Morris, M. (2001). *A Handbook for Value Chain Research*. IDRC, Ottawa, p. 133.

Lescuyer, G. and Nasi, R. (2016). Financial and economic values of bushmeat in rural and urban livelihoods in Cameroon: inputs to the development of public policy. *International Forestry Review*, 18, 93–107. https://doi.org/10.1505/146554816819683726.

MacDonald, D. W., Johnson, P. J., Albrechtsen, L., Seymour, S., Dupain, J., Hall, A. and Fa, J. E. (2012). Bushmeat trade in the Cross-Sanaga rivers region: evidence for the importance of protected areas. *Biological Conservation*, 147, 107–114.

Masse, F., Lunstrum, E. and Holterman, D. (2018). Linking green militarization and critical military studies. *Critical Military Studies*, 4, 201–221. https://doi.org/10.1080/23337486.2017.1412925.

Milliken, T. (2014). Illegal Trade in Ivory and Rhino Horn: An Assessment Report to Improve Law Enforcement under the Wildlife TRAPS Project. USAID and TRAFFIC. ISBN 978-1-85850-373-8.

MINADER (Ministry of Agriculture and Rural Development). (2015). The State of Biodiversity for Food and Agriculture in the Republic of Cameroon. Country Report, p. 219.

MINEPRD (Ministry of Economy, Planning and Regional Development). (2020). République of Cameroun, National Development Strategy 2020–2030, p. 244. ISBN: 978-9956-26-086-7.

Mongabay. (2012). New Reports from Inside Cameroon Confirm Grisly Mass Killing of Elephants. Retrieved 11 October, 2020, from https://news.mongabay.com/2012/03/new-reports-from-inside-cameroon-confirm-grisly-mass-killing-of-elephants-warning-graphic-photos/.

Strobel, S. (2019). Towards a Women-Oriented Approach to Countering Wildlife, Forest and Fisheries Crime (WFFC). UW-Platteville, USA.

Taylor, G., Scharlemann, J. P. W., Rowcliffe, M., Kümpel, N., Harfoot, M. B. J., Fa, J. E., et al. (2015). Synthesising bushmeat research effort in West and Central Africa: a new regional database. *Biological Conservation*, 181, 199–205.

Tchouto, M. G. P., Yemefack, M., De Boer, W. F., De Wilde, J. J. F. E., Van Der Maesen, L. J. G. and Cleef, A. M. (2006). Biodiversity hotspots and conservation priorities in the Campo-Ma'an rain forests, Cameroon. *Biodiversity and Conservation*, 15, 1219–1252. https://doi.org/10.1007/s10531-005-0768-6.

Tedou, J., She, E. J., G. B., Abanda A., Tatsinkou, C., Tchamago, K. O., Tchomthe, S., Kana, K. C., et al. (2011). Statistical directory of Cameroon. (In French) 2011 Edition, Yaoundé, p. 456.

United Nations Women. (2021). *Gender-Inclusive Peace Processes: Strengthening Women's Meaningful Participation through Constituency Building*. United Nations Peace, Security and Humanitarian Action Section. UN Women, New York.

Van Vliet, N. and Mbazza, P. (2011). Recognizing the multiple reasons for bushmeat consumption in urban areas: a necessary step toward the sustainable use of wildlife for food in Central Africa. *Human Dimensions of Wildlife*, 16, 45–54. https://doi.org/10.1080/10871209.2010.523924.

Van Vliet, N., Moreno, J., Gómez, J., Zhou, W., Fa, J. E., Golden, C., Alves, R. R. N. and Nasi, R. (2017). Bushmeat and human health: assessing the evidence in tropical and sub-tropical forests. *Ethnobiology and Conservation*, 6, 3. https://doi.org/10.15451/ec2017-04-6.3-1-44.

Wairima, G. (2016). When Canines Protect Wildlife | African Wildlife Foundation. Retrieved 13 October, 2020, from https:// www.awf.org/blog/when-canines-protect-wildlife.

Watson, R. T., Dias, B., Ghmez, R., Heywood, V. H., Janetos, T., Red, W. V. and Ruark, G. (1995). *Global Biodiversity Assessment: Summary for Policy-Makers*. Published for the United Nations Environment Programme by Cambridge University Press, Cambridge, pp. vii–F 46.

Wilkie, D. S. and Carpenter, J. F. (1999). Bushmeat hunting in the Congo Basin: an assessment of impacts and options for mitigation. *Biodiversity and Conservation*, 8, 927–955. https://doi.org/10.1023/A:1008877309871.

World Bank. (2016). Republic of Cameroon: Priorities for Ending Poverty and Boosting Shared Prosperity. Systematic Country Diagnostic. World Bank. Report No. 103098-CM.

WBG (World Bank Group). (2019). https://www.worldbank.org/en/country/cameroon/overview. Accessed on 12 December 2020.

7 Women in the Rhino Poaching Conflict

A Feminist Political Ecology Analysis

Francis Massé, Nícia Givá, and Elizabeth Lunstrum

Feminist Approaches on Poaching Economies and Militarized Responses

Using a feminist political ecological framework, this chapter examines the gendered dynamics of commercial wildlife poaching (e.g., organized, professional, economically driven, integrated). A feminist political ecology (FPE) of poaching integrates key insights from three disciplines: FPE, feminist criminology, and green criminology. The main objective of this chapter is to introduce readers to these three perspectives and their intersections through a review of the literature, while also drawing on ethnographic fieldwork. It highlights how these literatures provide conceptual pillars for developing a feminist approach to poaching. At the core of an FPE of poaching is an acknowledgment that a robust understanding of poaching economies benefits from a comprehensive examination of local gender and household relationships, as well as women's experiences of poaching and militarized responses. Insights that emerge from such examinations have diverse implications as they speak to the drivers of poaching, the less than obvious social harms of poaching conflicts, and possibilities for addressing both.

The chapter provides a case study of rhino (*Diceros bicornis, Ceratotherium simum*) poaching in the Mozambican–South African borderlands.[1] Women who are the wives of rhino poachers are often overlooked in analyses and interventions that concern the illegal rhino horn trade. Through focused research on and with these women, unique information emerged and new questions were asked and answered about the dynamics of rhino poaching and its impacts. Specifically, we asked how can a feminist approach to gender and the household identify and characterize the ways in which women were involved in the rhino poaching economy? How were women and their households impacted by poaching and law enforcement responses to it? How can women be more empowered as agents of change to reduce poaching and its ecological and social harms? In helping to answer these questions, an FPE outlook advances understanding of poaching economies that is necessary to help develop more locally responsive, socially just, and environmentally sustainable approaches.

DOI: 10.4324/9781003121831-9

Political Ecology, Criminology, and Conservation Crime

Political ecology is an approach for thinking about how and why relations of power shape interactions between humans and the environment and the resulting social and ecological outcomes that emerge. Political ecology has long interrogated how conflict over natural resources emerges from changing governance conditions of land and resources that enclose both for the benefit of some at the expense of others (Robbins, 2011). Political ecologists use this contextual lens to ask and to answer questions about conservation as a practice that has worked and continues to work to (intentionally or unintentionally) re-structure how people who live near natural resources access and use land and natural resources, which includes wildlife (Brockington, 2002; West, 2006). The management of protected areas, which are designated locations for biodiversity management, has often led to the direct and indirect displacement of some of the world's most socio-economically, culturally, and otherwise vulnerable people. Subsequently, this creates additional layers of marginalization and alienation of these people from the land and resources they depend on for their livelihoods, culture, and well-being (West et al., 2006; Agrawal and Redford, 2009; Kabra 2019; Witter and Satterfield, 2019). The extant literature highlights how the processes for protecting biodiversity can engender resentment and even resistance to conservationists, conservation, and wildlife by communities near protected areas (ibid.); these were the exact outcomes that conservation was designed to avoid.

Political ecology analyses of conservation have been at the forefront of scientific discourse about the contemporary poaching crisis (Duffy et al., 2019). For example, scholars who have contributed to political ecological analyses have examined how the social, economic, political, and historical drivers of rhino poaching specifically are associated with the ways in which conservation influences local peoples' economic poverty, political marginalization, and increased vulnerability (Hübschle, 2017; Witter and Satterfield, 2019; Lunstrum and Givá, 2020). These drivers have been, and continue to be, exacerbated by historical and ongoing displacement of people from their land as they are converted to conservation areas and/or through the top-down and often militarized ways in which the Mozambican and South African states have responded to rhino poaching (Lunstrum, 2014; Annecke and Masubele, 2016; Ramutsindela, 2016). Scholars have concluded that conservation and militarized anti-poaching responses fuel a feedback loop of failed biodiversity conservation, human exclusion, and livelihood preservation (Duffy et al., 2015; Roe and Booker, 2019). Diagnosing shortcomings, weaknesses, and gaps in conservation can help recalibrate and delineate opportunities for improvement.

Although political ecology brings a critical analytical lens to systemic issues of conflict around conservation, critical criminology offers a different and precise set of analytical tools for characterizing processes of criminality, criminalization, and victimization related to crime and enforcement responses

(Brisman, 2019). Such specificity offers a window into how uneven processes of political–economic, social, and cultural power shape understandings and processes of criminality, criminalization, policing, and their various impacts (Michalowski, 1996; Carrington and Hogg, 2017; Carrington et al., 2019). Critical criminology also examines the underlying socio-economic drivers that lead individuals to engage in illegal activities (Brisman, 2019). Recent work in criminology focused on the Global South (Carrington et al., 2016; Carrington and Hogg, 2017) and rural areas (Carrington et al., 2014; DeKeseredy, 2015) enhances critical criminology thinking by advancing perspectives on the roles of colonialism and capitalism in disrupting and restructuring the economic, social, and cultural relations of the countryside and rural areas. The critical criminological lens is thus a complement to political ecology when applied to the case of commercial poaching (e.g., Gore, 2017). Where political ecology draws attention to and unpacks the legacies of dispossession and violence with regards to conservation, critical criminology pays much needed attention to "the political, economic, and cultural forces that shape the definition and character of crime, and that frame the public and academic discourse about how we might achieve justice" (Michalowski, 1996, p. 9).

Green criminology lies at the theoretical intersection of political ecology and critical criminology. Green criminology is dedicated to "the exploration and examination of causes and responses to 'ecological,' 'environmental,' or 'green' crimes, harms, and hazards" (Brisman and South, 2013, p. 2). These include pollution, deforestation, climate change, electronic waste, and wildlife crime, among others. In at least this regard, green criminology is relevant to wildlife poaching because the problem inflicts uneven impacts and is responded to in uneven ways (e.g., Wyatt, 2013; Gore, 2017). Green criminology shares political ecology's focus on how political–economic processes shape environmental and related social harms (Davies, 2014; Goyes and South, 2017). Key to critical and green criminology is delineating the problem of, and thus solutions for, poaching economies found in the study of victimization. Victimization focuses on who or what is affected or harmed negatively by illicit activities and economies; it also considers the processes by which victimization starts, continues, and ends. For example, Davies (2014) examined processes of indirect or secondary victimization in the case of aluminum smelting. Davies (2014) explored how individuals who were not involved directly in the activity were impacted or harmed negatively – thus victimized – by the activity itself *and* by law enforcement responses implemented to mitigate risk. This chapter anchors upon the concept of indirect victimization to interpret the negative socio-economic impacts that women experience as a result of poaching economies and militarized law enforcement responses designed to eliminate poaching. Women can be primary and/or secondary victims of poaching economies and militarized responses through the combination of their husband's (or sons, brothers, fathers) participation in poaching, and through the militarized enforcement responses directed toward their male family members. Detecting and interpreting these unseen impacts of victimization within the broader context of poaching economies and militarized responses are requisites for effective interventions.

Brought together, these three disciplines create new spaces for asking and answering questions about the processes that shape human–environment interactions and human behavioral responses to such interactions, which include illicit resource use, responses to these, and related social and ecological harms. Integrating the three disciplines creates a road map for scientific advancement about the centrality of gender and gendered relations as important variables in understanding participation in poaching economies, how they are responded to, and how each is experienced. Beyond implications for the "on-the-ground" practice of conservation, this integration has the potential to influence gaps in theory. Criminologists have similarly called for a deepening of feminist and gendered analysis with regards to environmental crime (Sollund, 2017; Lynch, 2018; Sollund, 2020). Three pillars of an *FPE of poaching* emerge by drawing on feminist approaches within each of political ecology, criminology, and green criminology (Table 7.1).

Table 7.1 Integrating feminist political ecology, feminist criminology, and feminist green criminology enables an intersectional approach that explores the feminist political ecology of poaching

Theoretical Approach	*Guiding Principles from the Literature*	*Illustrative References*
Feminist political ecology	Gender and gender relations as variables in shaping human–environment interactions, control over, and access to land and resources, and experiences of socio-ecological change. Understanding embodied, lived experiences of political–ecological processes and change across scale.	Rocheleau et al. (1996); Schroeder (1999); Elmhirst (2011); Nightingale (2011); Hovorka (2012); Mollett and Faria (2013); Harcourt and Nelson (2015); Sultana (2015), (2021)
Feminist criminology	Gender and gender relations as variables in shaping participation in illicit activities, experiences of victimization of illicit activities, and enforcement responses. Understanding embodied, lived experiences of illicit activities, deviant behavior, enforcement responses.	Chesney-lind (2006); Chesney-Lind and Morash (2013); Renzetti (2013); Carrington et al. (2014); DeKeseredy (2015)
Feminist green criminology	Gender and gender relations as variables in shaping participation in environmental crime, harm, enforcement responses, and experiences of victimization. Understanding embodied, lived experiences of environmental crime, harm, and enforcement responses. Illicit activities, deviant behavior, and enforcement responses.	Davies (2014); Sollund (2017); Lynch (2018); Sollund (2020)

Feminist Approaches for Understanding Poaching Economies and Responses

Feminist Political Ecology. FPE focuses on gender as "a critical variable" in shaping scholarly exploration of, and thus demystifying, human–environment interactions, socio-ecological change, and access to land, forests, water, and wildlife (Rocheleau et al., 1996, p. 4). A primary concern of FPE is how the processes of power that shape social and ecological outcomes produce differential outcomes, opportunities, and challenges across gendered lines. FPE considers how, in both urban and rural settings, local gender relations and broader structures of patriarchy interact with local political economies of resource use, access, and exposure to environmental injustices to determine outcomes for men and women in their everyday lives (Elmhirst, 2011; Mollett and Faria, 2013; Sultana, 2015, 2021). Applied to conservation, FPE has provided insight into how new rules and spatial dimensions of biodiversity protection intersect with local norms about access use and gender relations to impact men and women differently (Schroeder, 1999; Perry and Gillespie, 2019). It remains an unanswered question: what might a parallel study reveal about the gendered impacts of poaching economies and intensified law enforcement or other sector responses to poaching?

FPE does not seek to displace other markers of identity like race, ethnicity, class, political orientation, or religion. Rather, FPE takes an intersectional approach. This means it accounts for how gender and gendered structures, norms, and workings of power overlap and intersect with dynamics related to class, race, sexuality, and broader histories of governing land, resources, and bodies, among other factors (Sundberg, 2004; Nightingale, 2011; Hovorka, 2012; Mollett and Faria 2013; Sultana, 2021). FPE also considers nuances and complexities of how human–environment interactions and their consequences are embodied and experienced by individuals differently in everyday life (Elmhirst 2011; Mollett and Faria 2013; Sultana 2015; Harcourt and Nelson 2015; Sultana 2021). This amounts to moving analysis of processes of power that shape unequal social and ecological outcomes away from the more generalized and high-level category of "the community" and into the specific; that is, to consider what is happening at the household level and with individuals. This down-scaling of analysis to more fine-grained levels allows the researcher to contextualize what broader processes and changes in political–ecological and human–environmental relations (e.g., a new protected area, development of new poaching economies, militarized conservation responses) mean for people and their daily lives now and into the future. Without such fine-grain analysis, important differences in how women and men experience these processes across scale may be under-considered, ignored, or homogenized. In practice, this may mean that there is a lack of adequate resources and inappropriate policy interventions aimed at reducing harms and supporting marginalized groups.

Feminist Criminology. Feminist criminology similarly places "gender at the center and not periphery of criminological theorizing and research"

(Carrington et al., 2016, p. 10). This includes moving "the social construction of femininity and masculinity to a central place in critical criminology" (Michalowski, 1996, p. 14). Feminist analyses help to uncover "distinctively different gendered patterns of crime and violence" (Carrington, 2014, p. 2). Much like FPE, feminist criminology "is not merely about adding women onto the agenda" (DeKeseredy, 2015, p. 181), but rather considers how women and gendered relations function across and between all dimensions of inquiry. For example, patriarchy and related gender–power structures intersect with political–economic (e.g., structural unemployment) and cultural processes (e.g., masculinities) to become an "explanatory factor in the relationship between gender and criminality" (DeKeseredy, 2015, p. 160). Feminist criminologists also interrogate how gendered cultural norms and related performances of masculinity combine differently to influence participation in illicit activities (Chesney-Lind and Morash, 2013; Renzetti, 2013; Carrington et al., 2014). Accounting for structures of gendered norms and relations can thus help to explain how women and men experience crime, enforcement responses, and processes of victimization (Messerschmidt, 1986; Chesney-Lind, 2006; Davies, 2014; Messerschmidt, 2013; Lynch, 2018). This enables a more nuanced, precise, and potentially predictive perspective on offending and victimization, which include those related to poaching economies.

Feminist criminology emerged in part because of policy and programmatic silence regarding women's victimization. The operationalization of patriarchy as an explanatory factor by feminist criminologists has helped explain the disproportionate victimization of women in certain types of crimes, such as sexual assault (Ogle and Batton, 2009). Feminist criminology has helped to name "the types and dimensions of female victimization" (Chesney-Lind, 2006, p. 7), which in turn can invite broadened perspectives into decision-making. Deconstructing processes of victimization is particularly relevant to illicit activities like rhino poaching, where women are excluded from the act of poaching itself, yet are still exposed to poaching-related risks, particularly from the poaching economy and militarized enforcement responses.

As argued by green criminologists themselves, green criminology also requires a stronger incorporation of gender and women's perspectives (South, 2014; DeKeseredy, 2015). Recent work attempts to address this shortcoming through the use of feminist approaches in green criminology. Lynch (2018), for example, demonstrated the value of feminist (green) criminology by drawing attention to female victims of green crimes. Intentionally not viewing gender as an isolated variable, Lynch (2018) maintained an intersectional approach to highlight the importance of socio–economic class to understand differential processes of women's green victimization. More recently, Sollund (2020) examined the gendered dimensions of crime against non-human animals. She combined masculinist and ecofeminist theories to explore why perpetrators who were caught killing predators illegally in Norway were predominantly male (Sollund, 2017). Given its focus on specifically *green* crimes and processes of victimization, a feminist green criminology offers a significant advancement for science about conservation crimes, such as rhino poaching.

Integrating feminist green criminology with FPE broadens the aperture about dimensions of power and human behavior vis-à-vis illicit natural resource use and economies. It also encourages FPE to engage more directly with conceptual and empirical questions about the *how* and *why* of environment-related criminality, victimization, and enforcement responses, all of which are dynamics that characterize poaching economies. For example, closely paralleling much political ecological work, criminological scholarship on rural masculinities and rural patriarchy highlights how certain rural norms and cultural constructions that influence criminal activity are associated with and essentialize "the enduring qualities of landscape – as tough, rugged and stoic 'sons of the soil'" (Carrington et al., 2014, p. 466). These insights begin to shed light on intersections between gendered norms, constructions of nature and landscape, and ecologies of particular areas. The integration of these disciplines helps to nuance critical criminology's "tend[ency] to romanticize the delinquencies of working class masculinity as a mode of resistance to the social order" (Carrington et al., 2014, p. 466). To be sure, similar resistance-oriented dynamics in conservation account partially for the motivations of rhino poachers, but scholars are typically cautious in romanticizing poachers as any type of freedom fighter (Hübschle, 2017; Lunstrum et al., 2021).

Toward a Feminist Political Ecology of Poaching

An FPE of poaching can be built on three overlapping pillars that are drawn from the feminist scholarship within and beyond political ecology and criminology (Figure 7.1). The first pillar entails a focus on gendered relations among actors and in local political economies and how they map onto the poaching economy, analysis of poaching, and evaluation of policy interventions. This entails examining local gender norms, how they influence engagement in poaching, and how they shape how particular groups (e.g., women) experience the poaching economy and enforcement responses. The second pillar involves scaling down the analysis of the poaching economy and its impacts to the household and individual levels. This scalar adjustment enables understanding of the multidimensional experiences of daily life, which helps to humanize what might otherwise be abstract, detached processes and analyses of the causes and consequences of wildlife poaching. Directly accounting for the lived and embodied realities of relevant stakeholders results in more accurate problem definitions and more environmentally just programs. The third pillar leverages ideas that are advanced from the other two, drawing specific attention to how women specifically experience poaching economies, crime prevention, enforcement responses, and related short-term and long-term consequences across scales. These pillars are presented as a heuristic tool to motivate and to inspire feminist-oriented research into

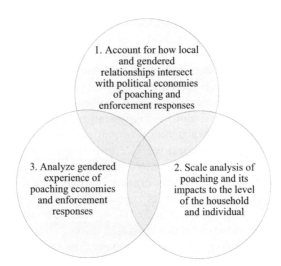

Figure 7.1 Three pillars of a feminist political ecology of poaching economies promote demarginalization of perspectives at different scales, from different lived experiences, and within multidimensional relationships.

poaching. Next, we present an illustrative example from the rhino poaching conflict in the Mozambique–South Africa borderlands.

An abbreviated context of rhino poaching in Mozambique and South Africa. Rhino poaching embodies many of the contemporary anxieties regarding commercial poaching and the illegal wildlife trade. These include threats to biodiversity, negative impacts on tourism economies, and broader fears around the destabilizing nature of transnational crime (Gore et al., 2019). The escalation of rhino poaching in South Africa, and specifically Kruger National Park, is more than an economy, however. It represents a conflict over a particular wildlife resource that involves (transnational) criminal syndicates, a violent poaching economy, and militarized and top-down responses by state and non-state actors aimed at curbing rhino deaths (Lunstrum et al., 2021). The causes and consequences of the rhino poaching economy in the Mozambique–South Africa borderlands, and efforts to address it, encapsulate the concerns that surround wildlife crime's meteoric spread and the intensification of securitized and militarized responses (e.g., Gore, 2017; Duffy et al., 2019).

Long-term research on poaching economies. The illustrative example used in this chapter draws on feminist research methods borne out through ethnographic research that was appropriate for capturing day-to-day realities of rhino poaching at the local level. Since 2003, one or more of the authors has collaborated as part of various long term research efforts in the Massingir region of Mozambique, adjacent to South Africa's Kruger National Park. The research focus on rhino poaching and related processes began in 2012, when the issue itself began to garner more attention by the conservation community. Collectively, the research effort has included hundreds of interviews with local residents, rangers, conservation personnel, the private and NGO sector, government officials, and policy makers. Each of the authors has also conducted separately, and collaboratively, participant and non-participant observation with local residents to understand their day-to-day realities. This longitudinal research has allowed for a synthesis of the many different aspects of rural life and the dynamics of conservation to more fully assess how different social, economic, political, and environmental issues and processes intersect, inform one another, and shape rhino poaching and its impacts.

The team's research focus on gender did not start with a focus on women. The dynamics of gender began to emerge as the authors asked study participants about why young men were participating in the poaching economy and what they perceived the impacts of the economy and militarized responses in the area to be. Newly purchased Toyota 4x4 trucks and freshly constructed concrete houses replaced the traditional mud and thatched-roof structures and no vehicles at all. These material goods signaled some of the locally realized benefits from the rhino horn economy. Some women and families have benefited economically from rhino poaching over both the short-term and long-term when men in their families returned successfully from rhino horn poaching trips. On the other hand, and representing the dark underside of the rhino poaching conflict, were the women and children whose husbands, fathers, brothers, and sons were killed or arrested while poaching. Stark contrasts began to emerge in local communities between households left more socio-economically vulnerable compared with those who were made more wealthy and resilient from the illegal rhino horn trade. At an individual level, some women and their families were ultimately rendered more vulnerable from rhino poaching, even if they were able to realize short-term material benefits. These observations and trends produced two new exploratory research questions: (1) what role(s) do women play in the current rhino poaching economy? and (2) how are women impacted by the rhino poaching economy and enforcement efforts? Below, these questions are addressed using the three pillars of an FPE of poaching.

Pillar 1. Local Gender Relations, Political Economies, and Women's Participation in the Rhino Poaching Economy

Although the above questions focus explicitly on women, long-standing local gender norms and relations, and the specific political–economic context of the area, are relevant. For example, overt expectations entail that men are the primary or sole income generator of a household and are responsible for materially providing for their wife/wives and children. In the natural resource-dependent areas of southern Mozambique adjacent to Kruger National Park, men have long fulfilled this role since colonial times through migrant labor. For generations, the majority of men from the area have traveled to South Africa, spending up to ten months a year working in the mines or citrus plantations. Traditionally, men sent remittances to their families and invested in cattle. Cattle have long been important signals of wealth in the region and are a form of resilience should hard times hit. The size and composition of a herd indicate material prosperity and also symbolize a man's power, success, and masculinity.

Masculinity is also closely tied to *lobola*. Lobola is a traditional Shangaan marriage system whereby a man and his family offer a 'payment' to a prospective wife's family as a gift for marrying their daughter (Furquim, 2016). Shangaan culture is polygamous, with many men marrying more than one wife and having children with each. The ability to lobola a wife and to provide for them and their children is a symbol of status and masculinity, each of which is highly valued within a community. Investing remittances and wages in cattle is a strategic behavior. As cattle reproduce, a man uses them for the lobola of a wife. Owning livestock and having enough material wealth to lobola a wife, and then being able to provide for the household, are what many men are expected to do to maintain his family's social, political, and cultural status within a community.

Women are largely responsible for domestic and subsistence agricultural labor, which includes caring for children, tending subsistence agricultural fields, fetching water and fuelwood, and caring for the sick. Women are often expected to marry at a young age, even as teenagers. Once married, women take on additional domestic responsibilities. One result is that women tend to obtain lower levels of education than men and have low levels of literacy (UNESCO, 2020). Gendered expectations for men and women underlay the implications of how the loss of a husband impacts his wife/wives, children, and their households. When a man is killed during a poaching expedition, his household(s) are often left more socio-economically vulnerable than before (see Pillar 2).

In the Mozambique–South Africa borderlands, women are not involved directly in the process of poaching a rhino. As far as we know, there has never been a publicly known case of a female poacher, and no women have been arrested for poaching rhino. Women are, however, involved indirectly in rhino poaching. Given women's overall status as being materially vulnerable and the fact that economic well-being can be secured in part or full by rhino poaching, it is not uncommon to find women, as with parents and other family members, encouraging men to enter into the rhino poaching economy (e.g., Sundström et al., 2019). How some women come to encourage certain men to participate in the poaching economy can be explained in part by gendered relationships and local political economies.

Diminishing opportunities for reliable wage–labor employment has translated into men's decreased ability to support families and an erosion of the traditional cycle of migrant workers sending remittances to buy cattle. Men noted repeatedly how reliable employment opportunities for Mozambicans in South Africa were fewer and fewer (Interviews 2012, 2013; also see De

Vletter, 2007). First, increased restrictions on international border crossings over the past decade make finding work more difficult. Today, passports are needed to legally cross at border points, but they are often unaffordable or inaccessible to young Shangaan men. Second, informal methods for border crossing have been curtailed; in the past, men walked across the unfenced international border through Kruger National Park, but increased securitization, law enforcement, and the militarization of the border since 2010 in response to rhino poaching has made such efforts more difficult and risky (Lunstrum, 2014; Annecke and Masubele, 2016; Ramutsindela, 2016). Shangaan men no longer walk across the international border into Kruger looking for migrant-based employment; the risk of getting arrested or shot is perceived as too high, further closing off possibilities of finding much needed work in South Africa (Interviews 2012, 2013). Third, some families have been displaced from Limpopo National Park while others in the park and adjacent nearby private reserves are subject to new restrictions on land and resource use. The result is an erosion of subsistence livestock raising and farming possibilities for affected households (Interviews 2012, 2016, 2019; also see Milgroom and Spierenburg, 2008; Witter and Satterfield, 2019; Lunstrum and Givá, 2020). Fourth, the region has endured recurring periods of drought since 2015, making resettlement and conservation-induced insecurity of agricultural livelihoods of many households even more challenging (Figure 7.2).

Although poverty motivates some individuals to enter the illegal rhino horn trade, it is in and of itself too broad and simplistic a phenomenon that ignores more nuanced dimensions (see Hübschle, 2017; Lunstrum et al., 2021). Poverty intersects dynamically with other social, economic, and cultural phenomena, which includes gendered relationships and norms of masculinity. Some men explained how the inability to earn enough money to invest in cattle, lobola a wife, and support one or more households left them feeling unable to fulfill socio-cultural and gendered expectations of supporting their household (Interviews 2013, 2017, 2019). Low levels of self-worth and morale among many men, especially young men, was a recurring theme in interviews about the poaching economy. For example, community leaders, men, and women all described times when women, whose husbands were struggling to support their family, saw the money made by their neighbor from a successful rhino poaching expedition and encouraged their husband to do the same, stressing how the poaching economy was perceived as helping to support family and fend off hunger (Interviews 2017, 2019). Several stories were recounted of neighbors or family members who, while during a period of drought and hunger, were asked why they were sitting around while their neighbors or friends were out "in the bush" (referring to poaching rhino) providing for their family (Interviews 2017, 2019). One woman described how she helped her husband hide the money he made from rhino poaching, which was up to 300,000 meticais (~$4,700 USD) per hunt (Interviews 2019). The money, she explained, helped the family substantially

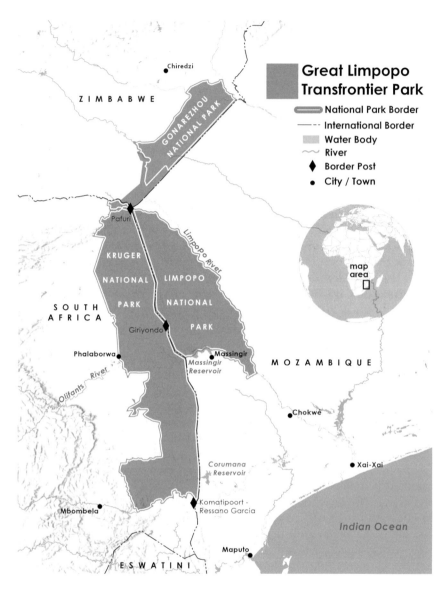

Figure 7.2 Much of the Great Limpopo Transfrontier Conservation Area (GLTFCA) overlaps with rhino poaching conflict at the Mozambique–South Africa borderlands.

while the man was alive. When her husband was killed by law enforcement authorities in Kruger National Park, the household was left with no income or source of material support. Given a long-term perspective and deepening poverty and insecurity in the region, it is not surprising some wives would encourage their husbands to hunt, even taking into account the grave risk.

The rhino poaching economy was perceived as a solution to these problems when recruiters offered opportunities to men to join poaching expeditions. Among study communities, it was evident the tension between the costs and benefits associated with the poaching economy was known to both men and women. Relations between men, women, and their families also provide less direct, even unintended encouragement for men to hunt. Given the high level of material poverty and lack of opportunities, some women actively seek out courtship with rhino poachers as it is perceived they can provide much-needed material security. The expectations around lobola have also changed with rhino poaching, now including more cash alongside cattle (Interviews 2019). So, even indirectly, and as a result of few other opportunities articulated with a more than reasonable desire for their own material well-being and security, women (and their families) play a subtle role in the dynamics of the rhino poaching economy.

Pillar 2. Scaling the Impacts of Rhino Poaching and Militarized Responses to Women and Households

Household-level perspective. The increase in rhino poaching in Kruger National Park has been met by a heavy-handed militarized response that has resulted in the deaths of hundreds of men (Lunstrum, 2014; Reuters, 2015; Annecke and Masubele, 2016). The impacts of militarized responses and poacher deaths on households, and particularly women, are however often obscured or under-examined by scholars and practitioners working on rhino poaching. Many women who lost their husband are left more vulnerable than they were previous to their husband's involvement in rhino poaching, even if they experienced short-term influxes of wealth from successful poaching expeditions. The emotional and material loss of a husband creates hardships for women left behind. However, simply pointing to the loss of a husband as creating hardship for women misses the nuanced ways in which gender relations come together with poaching and conservation's militarization to shape processes of women's indirect victimization.

First, the rhino poaching conflict is eroding the security and certainty enabled by lobola. Lobola provides social safety nets for families (Furquim, 2016). In an area with almost no opportunities for income or employment, lobola and marriage are viewed by parents as the primary means with which to gain material security for their daughters and for women to guarantee it for themselves. The rapid and abundant cash wealth derived from the poaching economy has resulted in poachers engaging in lobola and marrying an additional wife. An additional wife increases male cultural capital and allows poachers to assume traditional masculinized roles within their family and community. In this regard, marriage and wives are highly desirable from a socio-cultural perspective. However, because poaching is so lucrative compared with other means of employment, poachers are able to amass the wealth needed to lobola and marry rapidly and younger than normal. This resulted in men marrying at younger ages and taking younger wives.

Second, many women and their families viewed the wealth and associated socio-cultural security earned from rhino poaching as a potential way out of poverty and economic hardship for themselves and their children. This perspective is a double-edged sword as poaching increases the long-term vulnerability of women: rhino poaching-related wealth is unreliable because many of these young men are killed by anti-poaching forces in Kruger. Many widows had more of their lives ahead of them than behind when their poacher-husband was killed, which intensifies any negative effects. Cultural customs largely prohibit widows from re-marrying, which adds an additional layer of difficulty to find ways to provide for their family and to develop networks of resiliency. Confronted with militarized responses to poaching, what is traditionally a source of material security for households has become an unfortunate and unintended vicious cycle. Unfortunately, the region has long seen the premature death of men through dangerous labor in the South African mines, the Mozambican civil war, and more recently with the HIV/AIDS epidemic (e.g., Baltazar et al., 2015). Communities in Massingir are well aware that the rhino poaching economy causes men to die younger and at higher numbers (Interviews 2019). The result is an increasing number of women, especially young women, who are widowed and left without the material and emotional support of their husband; with some being married for only a short time, with young children to support, and lacking the experience to navigate household responsibilities alone.

Third, women, particularly mothers with children, are put involuntarily into a position where the primary income earner is unable to contribute, even during times when wage labor is possible. A lack of employment opportunities is coupled with low literacy and education levels among women to curtail wage–labor employment. A woman's cultural inability to re-marry after being widowed contributes to new hardships for women that have new and multiple roles to fill for their household. To have fields cleared and houses or granaries repaired or built, for example, widows must hire people from the community, which was difficult given the lack of income. Widows and their households were in a position of socio-economic-emotional precarity that they had not known previously.

Community-level perspective. Although the impacts of victimization and harm clearly manifest at the level of individual women and households, the impact and severity of rhino poaching and the militarized responses on women also accrue *across* households and communities. These group-level impacts crystalized when we "panned out" and recognized how the processes described above were happening across the many households, villages, and communities in and around Massingir and other towns in the Mozambican borderlands where the rhino poaching economy had taken hold. Hundreds, if not thousands, of men, women, and households have been affected. The primary fear and uncertainty that emerged when speaking with community members, and women specifically, was what the future held for them, their households, and their communities with the loss of so many young men (Interviews 2019).

How, they asked, do these households and their communities develop now and into the future without young men (Interviews 2019)? Rhino poaching and its militarized responses have intersected with and ruptured traditional gender relations and the structures built up around them in a way that has left an entire region more vulnerable. This is a point that Mozambican development NGOs and officials of the Mozambican government, such as in the Attorney General's office, have become increasingly concerned with (Interviews 2019).

Pillar 3. Women's Experience as a Foundation for Their Status as Change Agents

It is at the intersection of how women both suffer from and support men's engagement with the rhino poaching economy that their role as agents of positive change is recognized by the Mozambican government and NGOs (see Hübschle and Shearing, 2018, for more on agents of change to address rhino poaching). For example, one Mozambican NGO is working with women, and specifically widows, "to make noise about not poaching" (Interviews 2016). One way women were "making noise" was through radio shows that started in 2014. These radio shows feature women who speak about rhino poaching and the difficulties that they and their children have faced since their husbands' deaths. They do so in the hope that other women will hear the message and discourage their husbands from poaching. The programs also target men by having wives and mothers express how important and valuable young men are to families and community. Hence, these radio programs use the voices of women to spread the message to Mozambicans, "Don't go to Kruger, you will be killed!", as explained by a representative from the NGO. The same NGO representative explained how their messages are not ones of moral value about how it was wrong to kill a rhino. The messages are about spreading the simple message to men that they are likely to be killed if they go to Kruger, and they are a valuable and important member of their family, community, and country (Interviews 2016). Women, including those who once may have encouraged participation in the rhino poaching economy, are the central voices communicating these messages.

Advancing Feminist Perspectives on Poaching Economies

This chapter began by elaborating on FPE, feminist criminology, and green criminology as a way to focus on gender in the analysis of poaching. Synergies across and between these perspectives provided a foundation for an FPE of poaching. All three pillars of the approach were inspired by feminist approaches. The first concerned how local gender relations and gendered political economies intersect with political economies and ecologies of poaching. The second involved a re-scaling of analysis, which focused on how poaching, its impacts, and implications of enforcement responses manifest at the individual and household levels. The third was an explicit analysis of how women experience and are impacted by poaching economies and efforts to disrupt them.

The strength of the approach may be realized by future research on the gendered dynamics and implications of wildlife poaching or related conservation crime issues, such as wildlife trafficking, illegal fishing, or illegal logging. The exploratory analysis herein illustrated how local and long-standing gender norms have intersected with local political economies to shape women's role in rhino poaching, where some women encouraged their husbands to poach to provide an improved level of material security for their families. Many men who took the risk to poach rhino were killed by militarized responses to poaching, which has resulted in hundreds of widows in the Massingir area whose lives, households, and children are now more vulnerable. The changes in women's vulnerability and that of their households over both the short-term and long-term lay in how the lived experience of the death of a poacher intersects with long-standing gendered norms and structures by rupturing them and the social safety nets built up around them. Women are also agents of change in reducing rhino poaching and its harms. Women were recognized as having increasingly powerful voices in discouraging men from poaching by highlighting the value of men in their communities and the damage their loss to poaching caused for their families. The interdisciplinary feminist approach used may likely contribute to other research efforts and a more comprehensive understanding of poaching economies and militarized responses. Hopefully, this can result in the necessary base of evidence from which to develop more sustainable, just, and human-centered approaches to reduce the ecological and social harms of poaching.

Note

1 This chapter draws on similar theoretical framing and empirical material in a previously published article. Massé, F., Givá, N., & Lunstrum, E. (2021). A feminist political ecology of wildlife crime: The gendered dimensions of a poaching economy and its impacts in Southern Africa. *Geoforum*, *126*, 205–214. Permission has been granted for use of shared material. Please refer to the article for a complimentary discussion on gender relations of the rhino poaching economy and wildlife crime.

Literature Cited

Agrawal, A., & Redford, K. (2009). Conservation and displacement: an overview. *Conservation and Society*, *7*(1), 1–10.

Annecke, W., & Masubelele, M. (2016). A review of the impact of militarisation: the case of rhino poaching in Kruger National Park, South Africa. *Conservation and Society*, *14*(3), 195–204.

Baltazar, C. S., Horth, R., Inguane, C., Sathane, I., César, F., Ricardo, H.,... & Young, P. W. (2015). HIV prevalence and risk behaviors among Mozambicans working in South African mines. *AIDS and Behavior*, *19*(1), 59–67.

Brisman, A. (2019). Editor's Introduction to the special issue: "Crucial critical criminologies—Revisited and extended". *Critical Criminology*, *27*(1), 1–4.

Brisman, A., & South, N. (2013). Introduction: horizons, issues and relationships in green criminology. In *Routledge international handbook of green criminology* (pp. 17–40). Routledge.

Brockington, D. (2002). *Fortress conservation: the preservation of the Mkomazi Game Reserve, Tanzania*. Indiana University Press.

Carrington, K. (2014). *Feminism and global justice*. Routledge.

Carrington, K., Dixon, B., Fonseca, D., Goyes, D. R., Liu, J., & Zysman, D. (2019). Criminologies of the global south: Critical reflections. *Critical Criminology*, *27*(1), 163–189.

Carrington, K., Donnermeyer, J. F., & DeKeseredy, W. S. (2014). Intersectionality, rural criminology, and re-imaging the boundaries of critical criminology. *Critical Criminology*, *4*(22), 463–477.

Carrington, K., & Hogg, R. (2017). Deconstructing criminology's origin stories. *Asian Journal of Criminology*, *12*(3), 181–197.

Carrington, K., Hogg, R., & Sozzo, M. (2016). Southern criminology. *The British Journal of Criminology*, *56*(1), 1–20.

Chesney-Lind, M., & Morash, M. (2013). Transformative feminist criminology: a critical re-thinking of a discipline. *Critical Criminology*, *21*(3), 287–304.

Davies, P. A. (2014). Green crime and victimization: tensions between social and environmental justice. *Theoretical Criminology*, *18*(3), 300–316.

De Vletter, F. (2007). Migration and development in Mozambique: poverty, inequality and survival. *Development Southern Africa*, *24*(1), 137–153.

DeKeseredy, W. S. (2015). New directions in feminist understandings of rural crime. *Journal of Rural Studies*, *39*, 180–187.

Duffy, R., Massé, F., Smidt, E., Marijnen, E., Büscher, B., Verweijen, J.,... & Lunstrum, E. (2019). Why we must question the militarisation of conservation. *Biological Conservation*, *232*, 66–73.

Duffy, R., St John, F. A., Büscher, B., & Brockington, D. A. N. (2015). The militarization of anti-poaching: undermining long term goals? *Environmental Conservation*, *42*(4), 345–348.

Elmhirst, R. (2011). Introducing new feminist political ecologies. *Geoforum*, *42*(2), 129–132.

Furquim, F. M. (2016). A permanência do Lobolo e a organização social no Sul de Moçambique. *Revista Cantareira*, *25*, 5–15.

Gore, M. L. (Ed.). (2017). *Conservation criminology*. John Wiley & Sons.

Gore, M. L., Braszak, P., Brown, J., Cassey, P., Duffy, R., Fisher, J.,... & White, R. (2019). Transnational environmental crime threatens sustainable development. *Nature Sustainability*, *2*(9), 784–786.

Goyes, D. R., & South, N. (2017). Green criminology before 'green criminology': amnesia and absences. *Critical Criminology*, *25*(2), 165–181.

Harcourt, W., & Nelson, I. L. (Eds.). (2015). *Practicing feminist political ecologies: moving beyond the 'green economy'*. Zed Books Ltd.

Hovorka, A. J. (2012). Women/chickens vs. men/cattle: insights on gender–species intersectionality. *Geoforum*, *43*(4), 875–884.

Hübschle, A. M. (2017). The social economy of rhino poaching: of economic freedom fighters, professional hunters and marginalized local people. *Current Sociology*, *65*(3), 427–447.

Hübschle, A., & Shearing, C. (2018). Ending wildlife trafficking: local communities as change agents.The Global Initiative Against Transnational Organized Crime.

Kabra, A. (2019). Ecological critiques of exclusionary conservation. *Ecology, Economy and Society-the INSEE Journal, 2*(2354-2020-1298), 09–26.

Lunstrum, E. (2014). Green militarization: anti-poaching efforts and the spatial contours of Kruger National Park. *Annals of the Association of American Geographers, 104*(4), 816–832.

Lunstrum, E., & Givá, N. (2020). What drives commercial poaching? From poverty to economic inequality. *Biological Conservation, 245*, 108505.

Lunstrum, E., Givá, N., Massé, F., Mate, F., & Lopes-José, P. (2021). The rhino horn trade and radical inequality as environmental conflict. *Journal of Peasant Studies.* DOI: 10.1080/03066150.2021.1961130.

Lynch, M. J. (2018). Acknowledging female victims of green crimes: environmental exposure of women to industrial pollutants. *Feminist Criminology, 13*(4), 404–427.

Massé, F., Givá, N., & Lunstrum, E. (2021). A feminist political ecology of wildlife crime: the gendered dimensions of a poaching economy and its impacts in Southern Africa. *Geoforum, 126*, 205–214.

Messerschmidt, J. W. (1986). *Capitalism, patriarchy, and crime: toward a socialist feminist criminology*. Rowman & Littlefield.

Messerschmidt, J. W. (2013). *Crime as structured action: doing masculinities, race, class, sexuality, and crime*. Rowman & Littlefield.

Michalowski, R. J. (1996). Critical criminology and the critique of domination: the story of an intellectual movement. *Critical Criminology, 7*(1), 9–16.

Milgroom, J., & Spierenburg, M. (2008). Induced volition: resettlement from the Limpopo National Park, Mozambique. *Journal of Contemporary African Studies, 26*(4), 435–448.

Mollett, S., & Faria, C. (2013). Messing with gender in feminist political ecology. *Geoforum, 45*, 116–125.

Nightingale, A. J. (2011). Bounding difference: intersectionality and the material production of gender, caste, class and environment in Nepal. *Geoforum, 42*(2), 153–162.

Ogle, R. S., & Batton, C. (2009). Revisiting patriarchy: its conceptualization and operationalization in criminology. *Critical Criminology, 17*(3), 159–182.

Perry, N., & Gillespie, J. (2019). Restricting spatial lives? The gendered implications of conservation in Cambodia's protected wetlands. *Environment and Planning E: Nature and Space, 2*(1), 73–88.

Ramutsindela, M. (2016). Wildlife crime and state security in South (ern) Africa: an overview of developments. *Politikon, 43*(2), 159–171.

Renzetti, C. (2013). *Feminist criminology*. Routledge.

Reuters. (2015). Nearly 500 Mozambican poachers killed in S.Africa's Kruger since 2010-former leader. *Reuters.* Accessed Sept 26, 2016.

Robbins, P. (2011). *Political ecology: a critical introduction* (Vol. 16). John Wiley & Sons.

Rocheleau, D., Thomas-Slayter, B., & Wangari, E. (1996). A feminist political ecology perspective. *Feminist Political Ecology: Global Issues and Local Experience*, pp. 3–26. Routledge, London, UK.

Roe, D., & Booker, F. (2019). Engaging local communities in tackling illegal wildlife trade: a synthesis of approaches and lessons for best practice. *Conservation Science and Practice, 1*(5), e26.

Schroeder, R. A. (1999). *Shady practices*. University of California Press.

Sollund, R. (2017). Doing green, critical criminology with an auto-ethnographic, feminist approach. *Critical Criminology, 25*(2), 245–260.

Sollund, R. (2020). Wildlife crime: a crime of hegemonic masculinity? *Social Sciences, 9*(6), 93.

South, N. (2014). Green criminology: reflections, connections, horizons. *International Journal for Crime, Justice and Social Democracy, 3*(2), 5–20.

Sultana, F. (2015). Emotional political ecology. In Raymond L. Bryant (Ed.), *The international handbook of political ecology.* pp. 633–645. Edward Elgar Publishing.

Sultana, F. (2021). Political ecology 1: from margins to center. *Progress in Human Geography, 45*(1), 156–165.

Sundberg, J. (2004). Identities in the making: conservation, gender and race in the Maya Biosphere Reserve, Guatemala. *Gender, Place & Culture, 11*(1), 43–66.

Sundström, A., Amanda, L., Herbert N., Martin S., & Meredith L. G. (2020). Gender differences in poaching attitudes: Insights from communities in Mozambique, South Africa, and Zimbabwe living near the great Limpopo. *Conservation Letters, 13*, no. 1, e12686.

UNESCO. 2020. Mozambique. http://uis.unesco.org/en/country/mz. Accessed June 12, 2020.

West, P. (2006). *Conservation is our government now.* Duke University Press.

West, P., Igoe, J., & Brockington, D. (2006). Parks and peoples: the social impact of protected areas. *Annual Review of Anthropology, 35*, 251–277.

Witter, R., & Satterfield, T. (2019). Rhino poaching and the "slow violence" of conservation-related resettlement in Mozambique's Limpopo National Park. *Geoforum, 101*, 275–284.

Wyatt, T. (2013). *Wildlife trafficking: A deconstruction of the crime, the victims, and the offenders.* Springer.

8 Perceptions of Indigenous Baka Women's Inclusion in Wildlife Conservation and Exploitation

Jean Christian Mey Boudoug, Helen U. Agu, Pobo Kenfack Serge Ricardo, and Meredith L. Gore

Notions of Inclusion in Conservation

The Intergovernmental Science-Policy Platform on Biodiversity and Ecosystem Services (IPBES) explicitly recognized the detailed knowledge that indigenous people and local communities (IPLCs) possess about biodiversity. For example, Objective 3(b) of the 2019–2030 Workplan committed to "enhanced recognition of and work with indigenous and local knowledge systems." The IPBES Global Assessment's summary for policymakers recapped several principles and background related to IPLCs, which included "inclusion and participation of IPLCs, and women and girls in environmental governance and recognizing and respecting the knowledge, innovations, practices, institutions and values of IPLCs."

The notion of inclusion has been grossly underexplored in conservation (see Tulloch, 2020, for inclusion in conservation and ecology conferences or Smith et al. [2017] for diversity and inclusion in conservation networks). Furthermore, the notion of (workforce) diversity appears to be growing as a topic for consideration in conservation (see Poor et al., 2021, for supporting women as part of the workforce in conservation). In the field of organizational management, the term *inclusion* has evolved to replace *diversity*. Many groups have implemented diversity programs and instituted cultures of practice that pertain to diversity-related issues. For example, conservation practitioners' workshops nowadays integrate more of the "Inclusion–Diversity–Equity" dimension in their organizational structures to guarantee more committed, better satisfied, and better performing participants from diverse communities. Diversity is a dynamic concept and can be integrated into a group using a variety of targeted recruitment initiatives, education and training, career development, or mentoring programs. Inclusion represents a broader set of group norms and practices, which include employee participation, communication strategies, and community relations, for example. In this regard, the management of diversity involves inclusion.

Diversity is often defined as the various ways in which groups differ, but inclusion involves bringing together and harnessing different forces and resources in a way that is beneficial (Peters et al., 2017). It can involve observable

DOI: 10.4324/9781003121831-10

characteristics such as gender, race, ethnicity, and age and non-observable traits such as culture or cognition. Some scholars have suggested a solution by focusing on the advantages of employing members of different identity groups. However, diversity ignores the dynamics and consequences of exclusion (Prasad, 2001). Individuals from diverse social and cultural groups are often excluded from information networks and opportunities. Therefore, inclusion is not an "organizational commodity," but rather a person's ability to contribute fully and effectively to an organization. Many studies on diversity and inclusion outside the field of conservation focus on women and other minoritized individuals (e.g., Kossek & Zonia, 1993; Mor-Barak et al., 1998). This chapter contributes to the conservation literature by focusing on perceptions of indigenous women's inclusion in the global wildlife trade.

Indigenous People and Conservation in Cameroon

Cameroon's hunter-gatherers are part of a huge family of Indigenous Central African forest populations, which occupy a vast territory that extends west-to-east from the Congo Basin Forest to Lake Victoria. However, their numbers and actual distribution as a whole or in sub-groups were not known until recent times. Olivero et al. (2016) estimated that approximately 920,000 of these people occurred in the Central African forests, with >60% in Democratic Republic of the Congo. Together, Cameroon's hunter-gatherers and Mbororo pastoralists constitute the largest group of indigenous people in the country. IPLCs represent 0.4% of the total population of Cameroon; the three most well-known IPLC are the Bagyeli or Bakola (estimated at approximately 4,000 individuals), Baka (estimated at approximately 40,000 individuals), and Bedzan (estimated at approximately 300 people). Baka peoples live in the eastern and southern regions of Cameroon, Bakola and Bagyeli people live in an area in the south, and the Bedzan live in the central region (IWGIA, 2020; United Nations, 2007).

These indigenous peoples have extensive local knowledge systems (ILKs), which are adaptive knowledge systems accumulated over generations of social–ecological interactions in a localized context (Benyei et al., 2020; Berkes et al., 2000; Reyes-Garcia, 2015). ILKs include key skills, abilities, and practices to help manage human–nature and human–wildlife interactions effectively. ILK in Cameroon is reasonably explained by the fact that many indigenous peoples derive most of their livelihood directly from forested natural resource systems, such as hunting, gathering, and sustainable use of forested systems (Lueong, 2016; Ndameau, 2001; Pemunta, 2019). Although indigenous people are distributed across the forests of more than six countries, IPLCs in the Congo Basin forest have maintained similar traditional lifestyles and relationships with their sedentary neighbors (Olivero et al., 2016; Wodon et al., 2012). They continue to live in extreme poverty and score poorly on most development indicators. In addition, their daily life is reduced by the continuous and often increasing multipurpose use of their forests (Awuh, 2016; Boedhihartono et al., 2015; United Nations, 2007). The Baka people, for example, are dependent on wildlife for food and income (Hattori, 2005). Capacity building efforts have often occurred that used top-down, transfer-oriented models that employed intermediary or main farmer agents, which underlined the existing subordinate relationship between the Baka and neighboring farmers (Hattori, 2005).

Within this context of the dynamic relationship of indigenous people with the Congo Basin forest lies conservation. Some conservation projects and institutions in and around Cameroon's indigenous communities have worked to co-conserve biodiversity and to improve local livelihoods. Oftentimes, some conservation projects have attempted to sedentarize indigenous people through illegal appropriation of lands and threats to fundamental rights (Boedhihartono et al., 2015; Colchester, 2006; Colchester et al., 2006). Women were impacted disproportionately by these phenomena because they played a leading role in traditional hunting throughout Congo Basin forests using nets (Bailey & Aunger, 1985; Ichikawa, 1983; Terishima, 1983; Yasuoka, 2014). Unfortunately, Cameroon has experienced a continuous history of transitioning indigenous lands to conservation lands and for appropriating protected areas to private logging concessions (Boedhihartono et al., 2015; Cholez, 1999), although contemporary efforts have focused on securing indigenous people's rights to use and to manage natural resources. Nongovernmental organizations (NGOs) and government agencies have collaborated to adopt and to enforce international, national, and local mandates to improve local livelihoods (e.g., Boedhihartono et al., 2015; Endama et al., 2010; Sandker et al., 2009).

Today, multiple biodiversity conservation and efforts to preserve livelihoods have been implemented, with recurring quarrels between IPLCs and conservation actors. In 2016, the Rainforest Foundation UK, in its evaluation report on the governance of protected areas in the Congo Basin, raised the debate of the failure of conservation in the Congo Basin for both biodiversity and people. In its main conclusions, the report deplored the rise in violence and lack of respect for human (indigenous) rights within the implementation of conservation initiatives. Clear evidence was collected that incriminated many rangers and conservation NGOs as depriving ILPCs of their fundamental rights in 24 protected areas (Pyhälä et al., 2016). Some scholars speculated that the Baka lifestyle had not been considered sufficiently and, thus, that institutions had not shown much interest in community-based conservation (Hattori, 2005).

The role of indigenous women, such as the Baka, in Cameroonian forest conservation was not defined clearly by inclusive conservation frameworks. This is noteworthy because indigenous Baka women (IBW) were still responsible for most cooking, farming, and catering of the house and children. Hence, they had the potential to contribute substantial positive results for community-based conservation projects (Krietzman, 2019). Research from the Guatemala Maya Biosphere Reserve, for instance, suggested that women's particular roles and responsibilities within the household, community, and society led them to develop unique knowledge related to biodiversity; that is, women were in one of the most appropriate positions to bring about different perspectives and innovative solutions regarding biodiversity concerns (Palmer, 2018). Many scholars (Abrams et al., 2003; Graham et al., 2014; Lockwood, 2010) asserted that including indigenous women's voices more directly in conservation, promoting fairness during interactions with indigenous populations, and granting more direct engagement for decision-making processes will result in more sustainable and just governance of protected forest areas.

This chapter explores how inclusion of IBW was perceived by individuals who were external to the Baka community. This expert elicitation may be viewed as a primary step toward more complete understanding of IBW's roles in conservation (e.g., directly interviewing IBW about their own perceptions of inclusion). Expert elicitation is a well-known methodology in conservation; it is often used in conservation contexts where direct perceptions of individuals or groups cannot (or should not) be measured directly for a variety of reasons (Morss et al. 2018; Tobi & Kampen 2018; Viollaz et al. 2021). Experts can be defined in many ways; in this chapter, we define experts as individuals external to the Baka who were living and working in direct contact with them.

The Dja Biosphere Reserve

This study was conducted in the eastern and southern regions of Cameroon, in the Dja Biosphere Reserve (DBR) (Figure 8.1). DBR is a tropical lowland rainforest that was designated a UNESCO World Heritage Site in 1981. The reserve allowed human use of natural resources, and there was general consensus that the DBR has been degraded by human activities, which included wildlife poaching (Epanda et al., 2019) and also agriculture, such as coffee, cocoa, plantain, cassava, and groundnuts (Muchaal & Ngandjui, 1999). The DBR has a range of species within its borders, which include well-known species such as giant ground pangolins (*Smutsia temminckii*), chimpanzees (*Pan* spp.), gorillas (*Gorilla*), red duikers (*Cephalophus natalensis*), blue duikers (*Philantomba monticola*), elephants (*Loxodonta africana*), and brush-tailed porcupines (*Atherurus* spp.) (Muchaal & Ngandjui 1999). The administrative surface area of the DBR is about 526,004 ha (MINFOF and WRI, 2020). The human population of DBR is approximately 130,000, which is a density of 9.93 inhabitants/km^2 (BUCREP, 2005; MINFOF and WRI, 2020).

Indigenous Baka populations around the DBR reserve, especially those in the southern and eastern regions, total an estimated 40,000 individuals. They live in small groups usually close to roads or rivers and often in proximity to small settlements of non-Baka farmers. Women gather non-timber forest products (Boedhihartono et al., 2015; Bailey et al., 1992) and are also responsible for fishing, cultivating, saving the family money, and choosing housing for the family. Baka societies are matriarchal (Boedhihartono et al., 2015; Hewlett, 2014). When Cameroon launched Baka sedentarization programs in the 1960s, many moved toward the roads, but still kept their nomadic hunter-gatherer lifestyle (Boedhihartono et al., 2015; Bailey et al., 1992, 2009). Recent observations suggested that their traditional way of life was disappearing rapidly due to unequal debt-bondage with their non-indigenous neighbors; they were surviving by providing labor to farms and plantations (Boedhihartono et al., 2015). The Baka still carried out occasional expeditions into the forest for traditional hunting, but most of the time, they were involved as guides and employees for outsiders' commercial poaching efforts that were intended to supply bushmeat for more urbanized populations (Boedhihartono et al., 2015; Bailey et al., 1989; de Wasseige et al., 2010; van Vliet et al., 2010; Wilkie & Carpenter, 1999). Employment opportunities with outside commercial poachers provided an important source of revenue to purchase manufactured goods (Boedhihartono et al., 2015; Kitanishi, 2006).

Study Population and Sampling Frame

Data were collected from participants (n = 300) in villages (n = 10) out of 107 villages that surrounded the DBR in a buffer zone of about 10 km from its

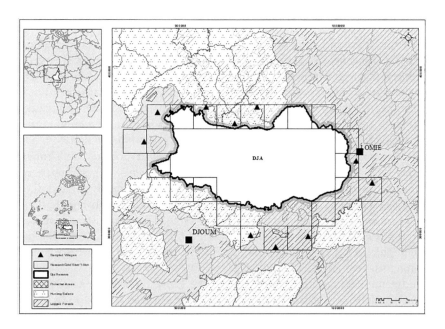

Figure 8.1 The Dja Biosphere Reserve in Cameroon is surrounded by several land use types, including protected areas with restricted use, hunting safari, villages, and agroforestry.

external boundaries from August to October 2020 using purposeful stratified sampling. Strata were derived based on a review of the geographic proximity of villages to the reserve, hunting safari camps, logged forests, and other types of timber management. The final stratification of villages was three in the north, three in the south, two in the northwest, and two in the northeast regions of the reserve. Individuals who self-identified as having regular experience with Baka people, with women, and with wildlife conservation were invited to participate. In practice, this meant that hunters, bushmeat sellers, rangers, researchers, students, and safari managers were included in the sample. Individuals <18 years of age, who self-identified as Baka or who were married to a Baka, were not able to participate in the interviews. During the study, verbal consent was given by every respondent each time, after general permission was given by the relevant administration.

The DBR is comprised of a complex mosaic of landscapes and land use types, and it was surrounded by logging, two types of agroforestry, and hunting zones as mandated by the 1994 Forestry Law of the Republic of Cameroon. The conservation environment was defined as a microlandscape, which characterized the various land use types within a 15 km^2 area within which selected villages occurred; we used publicly available satellite imagery analysis, country-wide land use system mapping data (1:200,000) (MINFOF and WRI, 2020), and forestry law classification types to define land use types. Based on these decision criteria, protected areas and hunting zones were the main land use types of interest for this research; logged forests and agroforestry areas were "other" types of relevant land use (Figure 8.1).

We studied ten sampled villages, which were distributed around the reserve and that comprised a variety of conservation and land use types (Figure 8.1). This included:

> (1) hunting zones/protected area/other (i.e., an area with an almost equal composition of protected areas, hunting zones, logged forests, and agroforestry spaces); (2) hunting zones/protected area (i.e., an area with an almost equal composition of protected areas and hunting zones); (3) hunting zones/other (i.e., an area with an almost equal composition of hunting zones, logged forests, and agroforestry spaces); and (4) other (i.e., an area of logged forests and/or agroforestry areas only).

Measurement

Perceptions of IBW's inclusion in conservation were measured using three four-point, Likert-type questions. An introductory question measured perceptions of indigenous women's inclusion in conservation in the DBR (i.e., "What do you think about IBW's inclusiveness in the Dja Biosphere Reserve?") followed by questions focused on perceptions of IBW's inclusion in wildlife conservation (i.e., "How do you think IBW are included in the bushmeat trade; ecotourism; anti-poaching; research and ecological surveys; hunting; hunting safaris?"). The next set of questions measured the different dimensions of perceived inclusion of IBW based on Arzumanyan and Mayrhofer (2016): (1) inclusion within the decision-maker's team; (2) inclusion within the permanent member team; (3) temporary inclusion as a team member for a particular work or task; and (4) exclusion from any of the groups. Inclusion within the decision-maker's team was measured using eight questions with scaled answers, where questions referred to the respondent's comfort with entrusting indigenous women within the framework of decision-making (e.g., we have IBW as permanent partners and/or assistant decision-makers in the management process of the DBR). Inclusion within the permanent member team was measured using the same approach, based on a respondent's comfort with engaging a Baka woman as a permanent member of his/her team or permanent business employee (e.g., we have IBW working with us as permanent staff). Temporary inclusion as a team member for a particular project or task referred to the respondent's ability to include a Baka woman in their activity as a temporary worker or participant; this was measured using the same approach (e.g., we temporarily employ IBW). Finally, exclusion from any of the above-mentioned circles was measured using the same approach and based on a respondent not including IBW in any of the above-mentioned ways (e.g., IBW are not part of the hunting safaris decision-making team).

Analytical Samples

Participants in the survey submitted willingly to a set of 30 questions related to their gender; age; origins/ethnicity; education level; employment status; and surrounding environment/micro-landscape type. They were invited to state their perceptions of indigenous women's inclusion in wildlife conservation and wildlife-related activities. Respondents who did not respond to all questions that were focused on inclusion were removed before analysis. A final sample of 282 interviews was analyzed. There were no statistically significant differences in demographic or outcome variables ($t = 0.01$, $p = 0.982$) between the full sample ($n = 300$) and the analysis sample ($n = 282$) (Table 8.1).

Analysis

Analysis was structured to increase understanding about IBW's inclusion in wildlife conservation. Mean scores for the perceived inclusion scale were calculated, and analysis of variance (ANOVA) was used to test the differences across conservation environments. Multiple linear regression was used to model the associated factors that were related to perceptions of IBW's inclusion. This test was appropriate to analyze hierarchical data in which respondents generated possible interdependent entry data. This method was

Table 8.1 Descriptive statistics of study participants (n = 300) from the Dja Biosphere Reserve

Variable		% (n)	Mean (SD)	% Missing (n)
Outcome variable				
Inclusion scale (1–4; more to less inclusive)			3.21 (0.31)	
Sociodemographic characteristics				
Gender	Male	63.33 (190)		
	Female	36.67 (110)		
	Total	100 (300)		
Age	18–23	0.67 (2)		
	24–29	23.33 (70)		
	30–35	41.33 (124)		
	35+	34.67 (104)		
	Total	100 (300)		
Origins	Local	83.33 (250)		
	Strangers	16.67 (50)		
	Total			
Activity	Rural	83.33 (250)		
	Conservation	16.67 (50)		
	Total	100 (300)		
Educational background	Below primary	13.33 (40)		
	Primary school graduate	36.67 (110)		
	High school graduate	33.33 (100)		
	University graduate	16.67 (50)		
	Total	100 (300)		
Micro-landscape type	PA/safari/other	22.00 (66)		
	PA/safari	21.33 (64)		
	Safari/other	9.33 (28)		
	PA/other	26.00 (78)		
	Other	21.33 (64)		
	Total	100 (300)		
Inclusiveness variable (1–4; more to less inclusive)				
Inclusion	Decision-maker		3.39 (0.60)	6 (18)
	Permanent participant		3.37 (0.61)	
	Temporary participant		3.07 (0.83)	
	Non-included		1.07 (0.92)	

inspired by Metheny and Stephenson (2019) to correct for downward bias in standard errors caused by non-independent data (Steele et al., 1996; Metheny & Stephenson, 2019), and it introduced error terms to allow for the possibility of unobserved heterogeneity (Diez-Roux, 2000; Metheny & Stephenson, 2019).

Results

Participants represented 120 families, which were spread across the 30 chosen villages. All had direct or indirect relationships with the DBR because the area covered a large landscape and was surrounded by different types of land use systems, but especially by forest management units for timber exploitation. Twenty-six percent of the respondents (n = 78) lived or had their active lives in an area co-dominated by both the DBR and a logged forest and/or an agroforestry area. Nine percent (n = 28) lived in an area co-dominated by hunting safaris and a logged forest and/or an agroforestry area.

The mean score on the scale of perceived inclusion of IBW in wildlife management and conservation was 3.21. The test of skewness indicated a statistically normal distribution (p = 0.634). The safari/other micro-landscape type was underrepresented, with two being identified compared to 12 forest management units for timber exploitation, and more than 20 community forest for timber and non-timber forest exploitation. Thirty-six percent of participants (n = 110) graduated from primary school, and 83% (n = 250) were engaged in private rural activities (Table 8.2).

Mean scores on inclusion did not differ significantly between micro-landscape. The interaction between micro-landscapes and inclusiveness was significant, which indicated that IBW from near forest management units, community forests, or agroforestry areas were perceived as more involved

Table 8.2 Mean inclusion scores for indigenous Baka women based on expert perceptions (n = 300)

Variable		Number of Respondents	Mean Inclusion Score (SD)	p-Value
Micro-landscape type	PA/safari/other	63	3.35 (0.07)	0.00
	PA/safari	59	3.14 (0.08)	
	Safari/other	26	3.27 (0.09)	
	PA/other	72	3.26 (0.06)	
	Other	62	2.42 (0.10)	
Gender	Masculine	179	3.09 (0.06)	0.56
	Feminine	103	3.04 (0.06)	
Age	18–23	2	3.00 (0.00)	0.95
	24–29	68	3.03 (0.08)	
	30–35	116	3.09 (0.07)	
	35+	96	3.08 (0.07)	
Origins	Local	234	3.08 (0.04)	0.75
	Strangers	48	3.04 (0.14)	
Activity	Rural	234	3.08 (0.04)	0.75
	Conservation	48	3.04 (0.14)	
Educational background	Below primary	35	3.09 (0.10)	0.53
	Primary school	103	3.15 (0.06)	
	High school	96	3.00 (0.07)	
	University	48	3.04 (0.14)	

Table 8.3 Interview respondents' perceived involvement of women in general and indigenous Baka women (IBW) specifically in various wildlife-related activities across different micro-landscape types adjacent to the Dja Biosphere Reserve (e.g., protected area [PA], safari, and other). Mean values and standard deviations are presented

Micro-Landscape Type	Perceived Involvement of Women in Wild Meat Trade	Perceived Involvement of IBW in Wild Meat Trade	Perceived Involvement of IBW in Ecotourism	Perceived Involvement of IBW in Anti-Poaching	Perceived Involvement of IBW Household
PA/safari/other	2.67 (0.19)	3.32 (0.06)[a]	1.79 (0.05)[a]	1.50 (0.06)[a]	1.50 (0.06)[a]
PA/safari	2.72 (0.16)	3.33 (0.02)[a]	1.73 (0.06)[a]	1.48 (0.06)[a]	1.48 (0.06)[a]
Safari/other	2.14 (0.33)[a]	3.39 (0.10)[a]	1.82 (0.07)[a]	1.75 (0.08)	1.95 (0.08)
PA/other	2.12 (0.19)[a]	3.27 (0.06)[a]	1.73 (0.05)[a]	1.42 (0.06)[a]	1.42 (0.06)[a]
Other	1.03 (0.18)	2.61 (0.02)[a]	1.80 (0.05)[a]	1.50 (0.06)[a]	1.50 (0.06)[a]

a p-Value < 0.05.

in wildlife-related activities (Xbar = 2.42) compared with those living in an environment under the direct influence of a protected area or hunting safaris.

Mean scores for inclusion perceptions also differed significantly between forest management units, community forests, agroforestry areas, and landscapes influenced by protected areas or hunting safaris. There was a significant positive perception of IBW's participation in the bushmeat trade (Xbar = 2.13). IBW were perceived as having temporary roles in the wild meat trade (Xbar = 3.34), involvement in ecotourism (Xbar = 1.50), and as paid workers (Xbar = 1.50). Interaction terms between inclusiveness and the other sociodemographic factors were not significant. Protected areas-based landscapes that were comprised of at least a protected area or a hunting safari tended to be exclusive in their inclusion of IBW in their management schemes (Table 8.3).

Discussion

This research explored expert perceptions of IBW's inclusion across a range of conservation-related behaviors in Cameroon. Sociodemographic variables were poor predictors of IBW's perceived inclusion in wildlife conservation in hunting safaris and protected areas. Micro-environment type variables helped to predict IBW's elevated positions in the illegal bushmeat trade near logged forests and agroforestry areas. This positionality can be situated within the context of low levels of local inclusion in resource supply chains and operational management.

It was not surprising that perceptions of IBW's ability to be included in conservation varied or that the most common perception of IBW's inclusion in decision-making was low overall. These are dominant projections and tropes. However, approaching inclusion and participation across a variety of micro-landscapes offered a range of new insights for governance of the DBR, particularly where the existing disparities were situated across protected area space. Perceptions of inclusion with micro-landscapes of logging forests were noteworthy, which portrayed Baka women as moderately active in wildlife activities as external facilitators. Perceptions of inclusion with the "other" micro-landscapes were stronger; that is, IBW were perceived as being permanently included in decision-making relevant to wildlife trafficking, and they were even thought to hold leadership roles in instances where only Baka communities were the focus (Xbar = 1.03). Importantly, when non-Baka women were involved in the illegal bushmeat trade, IBW were perceived to fulfill the roles of followers or facilitators.

Fa et al. (2016) explored organizations between Baka and non-Baka communities, which included men, and how the relationship impacted hunting, trading, and access to shotguns and rifles. The interdependency and dynamic relationship between Baka and non-Baka communities has a long history. Since the 1950s, the nomadic Baka hunter-gatherers have settled closer and closer to non-Baka/farmer villages, and the Baka have become more dependent on the farmers economically and other communities that were involved in logging, tourism, and conservation activities. Hattori (2014) reported the dependency created opportunities for indigenous inclusion, but it also presented conflicts of interest.

The micro-landscape types explored in this study helped offer a contextualization of how and potentially why IBW's inclusion was perceived the way it was. First, management of micro-environments was highly variable. Although funds for sustainable protection and conservation, surveillance, and monitoring activities were raised continuously for protected areas and hunting safaris, most other forest management units remained peripheral and less considered (WCS, 2014). Robinson (1999) suggested there was a strong correlation between forest logging and wildlife harvesting and trade within landscapes; they concluded that commercial logging contributed more than just changes to land management status (Yasuoka, 2006), but there was also local community pressure on local wildlife populations in the form of bushmeat provided from agroforestry areas (Boedhihartono et al., 2015). Poulsen et al. (2009) documented that development of industrial logging operations was associated with a 69% increase in the population of logging towns and a 64% increase in bushmeat supply in northern Congo between 2000 and 2006. Tieguhong & Zwolinski (2009) suggested that 62% of hunted animals were sold for cash income, and 38% were consumed by hunters and their families. Fa et al. (2003) observed that wild meat supply increased with proximity to protected areas, and MacDonald et al. (2012) observed that such proximity depended on the status of the protected area, with the most restrictive status being associated with smaller supplies. These studies explained the important bushmeat trade opportunities associated with logged forest micro-landscapes and IBW's perceived inclusion in the activity.

Changes in the micro-landscape economies have been linked to IBW's inclusion in different workforces. Gérard (1965) first documented progressive changes in Baka economy. His research assumed that Baka were moving progressively from nomadic to sedentary economies, which created an alternative structure to the traditional Baka economy that influenced IBW's roles in their communities significantly. These roles involved traditional Baka family's rulers and holders of the household economy; IBW were perceived as investing increasingly in food and revenue generation activities for their families. One change associated with this economic shift was their greater openness to trade and to exchange with Bantu non-Baka populations, especially women. What emerged was a larger and complementary organization that strengthened the bushmeat trade between male and female Baka as sources (hunters) and intermediate transporters (Baka women) to non-Baka women buyers around the DBR. The economic opportunities associated with agroforestry and agriculture micro-landscapes varied. IPLCs, which included Baka, gave up on their agricultural and agroforestry activities and invested more in the illegal wildlife trade because the latter was often perceived as being the fastest means to alleviate depressed economic status. Epanda et al. (2019) reported a significant difference in the frequency of illegal hunting between cocoa and non-cocoa producers; cocoa producers were less involved in poaching because they were involved in economically productive agroforestry.

The history of exclusion and expropriation for IPLCs within Cameroon's protected areas has been linked with colonial practices and post-independence conservation (Mkumbukwa, 2008); Baka communities have interacted with more sedentarized and educated Bantu communities. Although all communities may share the same riparian zone rights on paper, there were real and practical differences in the use of natural forest resources that created enduring forms of marginalization for the Baka. IBW continued to occur at the intersection of multiple forms of exclusion; they have almost always been excluded from key decision-making positions regarding management of natural resources in protected areas and hunting safari landscapes. IBW's roles in "logged and agroforestry" landscapes have been dominated by their responsibility to provide income for their family livelihood, regardless of their marital status (Homewood et al., 2020). These nuances justify a need to understand the role of IPLCs and women in wildlife conservation and exploitation more directly.

This research highlighted how social science interviews were able to produce novel insights about how indigenous women were perceived across a variety of conservation-related micro-landscapes, which illustrated variations in perceived involvement in wildlife conservation and exploitation. The theoretical concept of inclusion provided a useful platform for this research. Results helped to characterize differences in inclusion that depended on the living environment. Incorporating the detailed knowledge that IPLCs possess about biodiversity conservation and exploitation into decision-making does not have to be a difficult scientific or policy endeavor. A wealth of existing social science, conservation decision-making methods, and theoretical frameworks are available to incorporate IPLC voices, which include those of IBW.

Acknowledgment

Special thanks to the authorities and populations around the Dja Wildlife Reserve, in Djoum, Lomié, and Somalomo, for their consent, active contribution to the study, and for providing reliable information necessary for this research.

Literature Cited

Abrams, D., Rutland, A. and Cameron, L. (2003). The development of subjective group dynamics: Children's judgments of normative and deviant in-group and out-group individuals. *Child Development*, 74(6), 1840–1856.

Arzumanyan, L., & Mayrhofer, U. (2016). The adoption of digital tools in communities of practice-the case of group SEB. HAL.

Awuh, H. E. (2016). Access to discourse, marginalisation and exclusion in conservation-induced resettlement: The case of the displaced Baka of East Cameroon. *International Journal of Environmental Studies*, 73(2), 294–312. doi:10.1080/00207233.2016.1148446.

Bailey, R. C. (1985). *The sociology of Efe Pygmy Men in the Ituri Forest, Zaire (Africa, Behavior, Hunter-Gatherer)*. Cambridge, Massachusetts: Harvard University.

Bailey, R. C., Head, G., Jenike, M., Owen, B., Rechtman, R., Zechenter, E. (2009). Hunting and gathering in tropical rain forest: is it possible? *American Anthropologist*, 91, 59–82.

Bailey, R. C. & Aunger, R. Jr. (1989). Net hunters vs. archers: Variation in women's subsistence strategies in the Ituri forest. *Human Ecology*, 17(3), 273–297.

Bailey, R. C., Jenike, M. R., Ellison, P. T., Bentley, G. R., Harrigan, A. M. and Peacock, N. R., (1992). The ecology of birth seasonality among agriculturalists in central Africa. *Journal of Biosocial Science*, 24(3), 393–412.

Benyei, P., Arreola, G. and Reyes-García, V. (2020). Storing and sharing: A review of indigenous and local knowledge conservation initiatives. *Ambio*, 49(1), 218–230.

Berkes, F., Colding, J. and Folke, C. (2000). Rediscovery of traditional ecological knowledge as adaptive management. *Ecological applications*, 10(5), 1251–1262.

Boedhihartono, A. K., Endamana, D., Ruiz-Perez, M. and Sayer, J. (2015). Landscape scenarios visualized by Baka and Aka Pygmies in the Congo Basin. *International Journal of Sustainable Development & World Ecology*, 22(4), 279–291.

BUCREP, 2005. Repertoire actualise des villages du Cameroun: troisieme recensement general de la population et de l'habitat du Cameroun. http://www.ceped.org/ireda/inventaire/ressources/cmr-2005- rec_v4.7_repertoire_actualise_villages_cameroun.pdf

Cholez, C. (1999). Autonomie culturelle et autonomisation de la culture: Limites du regard anthropologique dans l'analyse des relations entre Pygmées et Grands Noirs. *Journal Des Anthropologues Association Française Des Anthropologues*, 77–78, 177–192.

De Wasseige, C., Devers, D., De Marcken, P., Eba'a Atyi, R., Nasi, R., Mayaux, P. (2010). *The Forests of the Congo Basin: State of the Forest 2008*. Luxembourg: Publications Office of the European Union.

Endamana, D., Boedhihartono, A., Bokoto, B., Defo, L., Eyebe, A., Ndikumagenge, C., Sayer, J. (2010). A framework for assessing conservation and development in a Congo Basin forest landscape. *Tropical Conservation Science*, 3, 262–281.

Epanda, M. A., Fotsing, A. J. M., Bacha, T., Frynta, D., Lens, L., Tchouamo, I. R., & Jef, D. (2019). Linking local people's perception of wildlife and conservation to livelihood and poaching alleviation: A case study of the Dja biosphere reserve, Cameroon. *Acta Oecologica*, 97, 42–48.

Fa, J. E., Currie, D., & Meeuwig, J. (2003). Bushmeat and food security in the Congo Basin: Linkages between wildlife and people's future. *Environmental Conservation*, 30(1), 71–78.

Fa, J. E., Olivero, J., Farfan, M. A., Lewis, J., Yasuko, H., Noss A, et al. (2016) Differences between Pygmy and non-pygmy hunting in Congo Basin forest. *PLoS ONE*, 11(9), e0161703. doi:10.1371/journal.pone.0161703.

Gérard, A. (1965). Changements sociaux chez les Pygmées Baka de l'Est-Cameroun. *Cahiers d'études africaines*, 5(20), 561–592. doi:10.3406/cea.1965.3050.

Graham, D., Wallace, V., Selway, D., Howe, E. and Kelly, T. (2014). Why are so many Indigenous Women Homeless in Far North and North West Queensland, Australia? Service Providers' Views of Causes. *Journal of Tropical Psychology*, 4. E8. doi:10.1017/jtp.2014.8

Hattori, S. (2005). Nature conservation and hunter gatherers' life in Cameroonian rainforest. *African Study Monographs. Supplementary Issue*, 29, 41–51.

Hattori, S. (2014). Current issues facing the forest people in southeastern Cameroon: The dynamics of Baka life and their ethnic relationship with farmers. *African Study Monographs*, 47 (Supp), 97–119, March 2014.

Herzog, S. (2013). Wildlife management i protected areas-goals and concepts. Conference Paper, 5th Symposium for Research in Protected Areas, pp. 295–298.

Hewlett, B. S. (2014). *Hunter-Gatherers of the Congo Basin: Cultures, Histories, and Biology of African Pygmies*. New Brunswick (NJ): Transaction Publishers.

Homewood, K., Nielsen, M. R., & Keane, A. (2020). Women, wellbeing and wildlife management areas in Tanzania. *The Journal of Peasant Studies*, 1–28. doi:10.1080/03066150.2020.1726323.

Ichikawa, M. (1983). An examination of the hunting-dependent life of the Mbuti Pygmies, east Zaire. *African Study Monographs*, 4, 55–76.

International Work Group for Indigenous Affairs (IWGIA). (2020). The Indigenous World Day 2020, report, Copenhagen, Denmark, pp. 3–4.

Kitanishi, K. (2006). The impact of cash and commoditization on the Baka hunter-gatherer society in southeastern Cameroon. *African Study Monographs Supplementary Issue*. 33, 121–142.

Kossek, E. E. and Zonia, S. C. (1993). Assessing diversity climate: A field study of reactions to employer efforts to promote diversity. *Journal of Organizational Behavior*, 14(1), 61–81.

Krietzman, R. (2019). Women in Conservation: A Study of Effective Community-Based Conservation and the Empowerment of Women in Tanzania. Independent Study Project (ISP) Collection. 3028. https://digitalcollections.sit.edu/isp_collection/3028

Lockwood, M. (2010). Good governance for terrestrial protected areas: A framework, principles and performance outcomes. *Journal of Environmental Management*, 91(3), 754–766.

Lueong, G. M. (2016). *The forest people without a forest: development paradoxes, belonging and participation of the Baka in East Cameroon*. New York, Berghahn Books.

Macdonald, D. W., Johnson, P. J., Albrechtsen, L., Seymour, S., Dupain, J., Hall, A. and Fa, J. E. (2012). Bushmeat trade in the Cross–Sanaga rivers region: Evidence for the importance of protected areas. *Biological Conservation*, 147(1), 107–114.

Metheny, N., & Stephenson, R. (2019). Political environment and perceptions of social inclusion after nationwide marriage equality among partnered men who have sex with men in the USA. *Sexuality Research and Social Policy*, 16, 521–528. doi:10.1007/s13178-018-0357-6.

Ministry of Forestry and Wildlife (MINFOF) and World Resource Institute (WRI). (2020). Atlas Forestier Interactif du Cameroun. https://www.wri.org/data/interactive-forest-atlas-cameroonatlas-forestier-du-cameroun

Mkumbukwa, A. R. (2008). The evolution of wildlife conservation policies in Tanzania during the colonial and post-independence periods. *Development Southern Africa*, 25(5), 589–600. doi:10.1080/03768350802447875.

Mor-Barak, M. E. and Cherin, D. A. (1998). A tool to expand organizational understanding of workforce diversity: Exploring a measure of inclusion-exclusion. *Administration in Social Work*, 22(1), 47–64.

Morss, R., Lazrus, H., & Demuth, J. (2018). The "inter" within interdisciplinary research: Strategies for building integration across fields. *Risk Analysis*. doi:10.1111/risa.13246.

Muchaal, P. K. & Ngandjui, G. (1999). Impact of village hunting on wildlife populations in the western Dja Reserve, Cameroon. *Conservation Biology*, 13(2), 385–396.

Ndameau, B., 2001. Protected areas and Indigenous peoples: the paradox of conservation and survival of the Baka in Moloundou region (south-east Cameroon). Case Study 7: Cameroon-Boumba Bek.

Olivero, J., Fa, J. E., Farfán, M. A., Lewis, J., Hewlett, B., Breuer, T, Carpaneto, G.M., Fernández, M., Germi, F., Hattori, S. and Head, J. (2016). Distribution and numbers of pygmies in Central African Forests. *PLoS ONE*, 11(1), e0144499. doi:10.1371/journal.pone.0144499.

Palmer, C. P. (2018). The role, influence and impact of women in biodiversity conservation. International Institute for Environment and Development (IIED), Public post on iied.org consulted on November 15th 2020.

Pemunta, N. V. (2019). Fortress conservation, wildlife legislation and the Baka Pygmies of southeast Cameroon. *GeoJournal*, 84(4), 1035–1055.

Peters, C., Cellucci, L., & Eric W. (2017). Diversity and inclusion within the Journal of Case Studies. *Journal of Case Studies*, 35(1), 1–10.

Poor, E. E., Imron, M. A., Novalina, R., Shaffer, L. J., & Mullinax, J. M. (2021). Increasing diversity to save biodiversity: Rising to the challenge and supporting Indonesian women in conservation. *Conservation Science and Practice*, 3(6), e395.

Poulsen, J. R., Clark, C. J., Mavah, G. & Elkan, P. W. (2009). Bushmeat supply and consumption in a tropical logging concession in northern Congo. *Conservation Biology*, 23(6), 1597–1608.

Prasad, A., 2001. Understanding workplace empowerment as inclusion: A historical investigation of the discourse of difference in the United States. *The Journal of Applied Behavioral Science*, 37(1), 51–69.

Prins, H. H. T., Grootenhuis, J. G., & Dolan, T. T. (Eds.). (2000). *Wildlife Conservation by Sustainable Use*. doi:10.1007/978–94-011-4012-6.

Pyhälä, A. A., Osuna Oroza, A., & Counsell, S. (2016). Protected Areas in the Congo Basin: Failing both People and Biodiversity? Rainforest Foundation UK. http://blog.mappingforrights.org/wp-content/uploads/38342-Rainforest-Foundation-Conservation-Study-Web-ready-embarg.pdf

Reyes-García, V. 2015. The values of traditional ecological knowledge, In: Martínez-Alier, J. & Muradian, R. editors. *Handbook of Ecological Economics*. Cheltenham: Edward Elgar Publishing, pp. 283–306.

Robinson, J. G., Redford, K. H., & Bennett, E. L. (1999). Wildlife harvest in logged tropical forests. *Science* 284(5414), 595–596.

Sandker, M., Campbell, B. M., Nzooh, Z., Sunderland, T., Amougou, V., Defo, L., Sayer, J. (2009). Exploring the effectiveness of integrated conservation and development interventions in a Central African forest landscape. *Biodiversity Conservation*, 18, 2875–2892.

Smith, N. S., Côté, I. M., Martinez-Estevez, L., Hind-Ozan, E. J., Quiros, A. L., Johnson, N., Green, S. J., Cornick, L., Shiffman, D., Malpica-Cruz, L. & Gleason Besch, A., (2017). Diversity and inclusion in conservation: a proposal for a marine diversity network. *Frontiers in Marine Science*, 4, 234.

Steele, F., Diamond, I., & Amin, S. (1996). Immunization uptake in rural Bangladesh: A multilevel analysis. *Journal of the Royal Statistical Society. Series A: Statistics in Society*, 159(2), 289–299.

Terishima, H. (1983). Mota and other hunting activities of the Mbuti archers: a socio-ecological study of subsistence technology. *African Study Monographs*, 3, 71–85.

Tieguhong, J. C., & Zwolinski, J. (2009). Supplies of bushmeat for livelihoods in logging towns in the Congo Basin. *Journal of Horticulture and Forestry*, 1(5), 065–080. Available online http://www.academicjournals.org/jhf.

Tobi, H., & Kampen, J. K. (2018). Research design: The methodology for interdisciplinary research framework. *Quality & Quantity*, 52, 1209–1225.

Tulloch, A. I. (2020). Improving sex and gender identity equity and inclusion at conservation and ecology conferences. *Nature Ecology & Evolution*, 4(10), 1311–1320.

United Nations. D.o.t.R.o.I.P. 2007. *United Nations Declaration on the Rights of Indigenous Peoples*. New York, United Nations Department of Public Information.

Van Vliet, N., Milner-Gulland, E., Bousquet, F., Saqalli, M., Nasi, R. (2010). Effect of small-scale heterogeneity of prey and hunter distributions on the sustainability of bushmeat hunting. *Conservation Biology*, 24, 1327–1337.

Viollaz, J., Long, B., Trung, C. T., Kempinski, J., Rawson, B. M., Quang, H. X., Hiền, N. N., Liên, N. T. B., Dũng, C. T., Huyền, H. T., & McWhirter, R. (2021). Using crime script analysis to understand wildlife poaching in Vietnam. *Ambio*, 50(7), 1378–1393.

Wildlife Conservation Society (WCS). (2014). Suivi de la Gestion de la Faune dans les Concessions Forestières au Cameroun. Document de Project. WCS-Cameroon. 12P.

Wilkie, D.S. & Carpenter, J.F. (1999). Bushmeat hunting in the Congo Basin: an assessment of impacts and options for mitigation. *Biodiversity and Conservation*, 8, 927–955.

Wodon, Q., Backiny-Yetna, P., & Ben-Achour, A. (2012). Central Africa: The case of the pygmies. In: Hall, G., & Patrinos, H., editors. *Indigenous Peoples, Poverty and Development*. Cambridge: Cambridge University Press, pp. 118–148.

Yasuoka, H., 2014. Snare hunting among Baka hunter-gatherers: implications for sustainable wildlife management. *African Study Monographs. Supplementary issue*, 49, 115–136.

Yasuoka, H. (2006). The sustainability of duiker (cephalophus spp.) hunting for the baka hunter-gatherers in southeastern Cameroon. *African Study Monographs, 33* (Supp): 95–120, May 2006.

Box 3

Voices from the Field

Influencers

Helen U. Agu and Meredith L. Gore

Individuals can link or connect to wildlife trafficking via a range of relationships, connections, and networks; these linkages have varying strength and longevity. Women may influence individuals or groups of people involved in wildlife trafficking; influence can be both positive and negative and be directed to other women or men. Family members, religious figures, educators, friends, and romantic partners may all be *influencers*, as can be female social media influencers on Instagram or Weibo. When we asked experts their perceptions of how, when, where, and why women are motivated to influence wildlife trafficking (study design, methods, and analysis discussed in Agu et al.[1,2] decision making power was repeatedly mentioned, particularly inside the household, as these voices from the field help illustrate:

> I don't know if women understand that they have the power to stop it. They're the mothers of the traffickers. They are the sisters that is why they have the power to stop it. I think it is the most important question there is. Do they understand that they have the power to stop it?
> (Female 1.1.4)

> A woman is in charge of many, many decisions in the life of her children in the life of her husband, and so on, and therefore I don't see or understand why when it comes to wildlife trafficking, it almost seems like an unexplored area entirely, you know. And this, I think, is very symptomatic of women in general being overlooked as key stakeholders globally, not just in wildlife trafficking, but in everything. And this is a very crucial problem in our societies. I think you can't just look at the women's participation because like I said, they rarely are in the lead position. And also women can be the very answer to resolving the problem. I think if we engage women at the village level, at the very lowest level, even to understand how we can really convince their sons, their brothers, their fathers to not commit these crimes and how to be on the right side of the law, women, in fact can be very strong advocates in those

societal structures, especially in Africa, where it's very hierarchical, and a woman, you know, she kind of manages a lot of things.

(Female 1.3.5)

The formula of the family is that they have the power to create a very good awareness thinking or positive side to the family so [the family] knows to do it or not to… so children are listening to their moms more than they do to their elders or whatever.

(Male 1.3.1)

Leader in Africa, the world, many of the inspirational leaders are women… part of me wants to say that it's the knowledge of the experts, and these women champions who can make a huge difference.

(Male 1.5.3)

Beyond decision making power, experts linked influencers to identities and opportunities, as these voices from the field help illustrate:

Being identified as a guilty person is very critical… sometimes a woman might have a chance to lose [their] job because of stigma from wildlife trafficking.

(Male 1.3.2)

With this mother's hand, no poached animal product the end of the day…comes to the household.

(Female 1.3.2)

Women influence the commission of wildlife crime. So, for example, we men may want meat for the children, or income. Will they influence (*change*) their stance or we are supposed to go out and commit the crimes?

(Female 1.3.2)

I do know there are many households in [REDACTED] where they keep the cheetah, the cheetah cubs can be transported, until they get the call the women feed them and take care of them. Some of the cheetahs will die from poor conditions but the women will try and be responsible to take care of them until they are moved to [REDACTED].

(Male 1.4.6)

Some experts recognized the broader context within which wildlife trafficking occurs, linking the act of influence to socio-environmental conditions:

…women pressure their husbands to…get involved in the poaching economy along the supply chain. And that's in part because you're in an area that has extremely high levels of, you know, poverty, drought and hardship.

(Male 1.4.5)

...a lot of these enterprises are done in the village for the source and even among the transfer routes. They go through people's homes. And so it is more than likely that the family is aware of what's happening and perhaps even complicit in it.

(Male 1.5.3)

Voices from the field inspire interesting questions about the role of women as influencers in wildlife trafficking. It is unclear if women in the roles of influencers are motivated by opportunity, marginalization, or masculinization. What boundaries exist between influencers and offenders, and how permeable are those boundaries in certain conditions? Are particularly influencer roles associated with certain forms of social control and are some forms of influencer more durable than others? Answers to these and other questions may help reduce harms associated with wildlife trafficking.

Notes

1 Agu, H.U., Andrew, C., and M. L. Gore. (2021). Mapping Terra Incognita: An Expert Elicitation Of Women's Roles in Wildlife Trafficking. *Frontiers in Conservation Science*. https://doi.org/10.3389/fcosc.2021.683979
2 The Michigan State University Human Subjects Protection Program approved the methods and analysis for the study (STUDY00003659) as exempt under 45 CFR 46.104(d) 2(ii).

9 Shaping Life in the Shadows

Gendered Dimensions of Wildlife Economies and Interventions in Central African Republic

Carolyn A. Jost Robinson, Elizabeth S. Hall, and Melissa J. Remis

The Underexplored Roles of Women in Wildlife Economies

The lack of scholarly attention to the variety of roles women play in contemporary wildlife economies (e.g., ecotourism, trophy hunting,) across economic, nutritional, and social dimensions may have roots in the poorly understood role of women in subsistence hunting. In anthropology, male-centric studies of subsistence hunting were pervasive through the late 20th century until feminist scholars shifted focus to include the dynamic importance of women's roles in resource use and subsistence patterns (e.g., Dahlberg, 1981; Leacock, 1978). Despite these ethnographic studies that documented the integral roles women have served in hunting economies from at least the 1960s forward, gendered narratives remain more common in natural resource studies; one result is that women remain marginalized in wildlife conservation initiatives (Agarwal, 2009; O'Shaughnessy & Krogman, 2011).

Thus, today, the knowledge base that underlies interventions to strengthen legal wildlife economies focuses overwhelmingly on wildlife populations, commodity chains, and law enforcement. These focal points are often male-centered, can be gender-biased in their derivation, and coupled with additional sources of gender bias in policymaking that shape interventions that either overlook or conflate women's roles in wildlife economies (Gore & Kahler, 2012). The research that addressed diversity in female and male roles in legal and illegal wildlife economies emphasized gendered differences in resource use rather than the complementary spaces from which they emerged (e.g., Lowassa et al., 2012; O'Shaughnessy & Krogman, 2011). These reductionist studies have value to be sure; however, they disaggregated essential information about content from context and consequence. Feminist political ecology challenges scholars to move beyond reducing women to a monolithic group and to allow the deep and diverse complexities of women's relationships with the environment to emerge more clearly. The work of feminist political ecologists emphasized that lived experiences of people were not homogenous and that these experiences were not static across time (Agarwal, 1992). Thus, there is a gap in any holistic assessment on the variety of ways that women navigate their essential economic and social roles in a dynamic wildlife economy.

Using an Ethnographic Approach to Describe Life in the Shadows

This research applied a multidisciplinary anthropological lens to understand women's choices, not only in the context of legal and illegal wildlife economies, but also within the broader context of continual socioeconomic and cultural shifts across time. This assessment was based on long-term research with hunter-gatherers and agriculturalists in the Dzanga-Sangha Protected Area (DSPA), Central African Republic (CAR). Ethnography and nutritional analyses enabled the examination of women's lived experiences in DSPA through the retelling of daily life by women and through insights gleaned through the embodied stories told by nutritional markers (Jost Robinson & Remis, 2016; Remis & Jost Robinson, 2014). The ethnography engaged the ingenuity and resourcefulness of local communities in DSPA to inform and to improve future interventions in illegal wildlife economies without further disenfranchising women and other marginalized populations who were dependent on natural resources. Building on the work of Lowassa et al. (2012), who examined gender complementarity within wildlife hunting economies in Tanzania and Ethiopia, this chapter explores the ways in which women of DSPA exercised autonomy within what were assumed explicitly to be complementary roles of production of tangible and intangible wealth within the household and broader community. The assumption that the productive roles of men and women were mutually exclusive was a common misconception that was replicated in livelihood interventions, particularly those from the field of wildlife conservation; the assumption resulted in exacerbating existing burdens faced by women in wildlife economies (Cookson, 2018).

This research employed the concept of "shaping life in the shadows" to help reorient our analysis for deeper understanding about the varied experiences for women of policy interventions and outcomes for local communities in DSPA. The use of shadows is not meant to discriminate gendered spaces; the term is used to delineate complementary spaces occupied by men and women within a society. A shadow is the dimensional space that takes shape when light is blocked by an object. In this example, we differentiated and described the often-obscured or unauthorized spaces in which women's lives took shape through their relationships to the broader wildlife economies, gendered dimensions of labor, and conservation interventions that intersected with daily life. Conservation anthropologists have a clear opportunity to acknowledge the ways that conservation interventions have overlooked issues specific to women—unbalanced interventions create shadows. In these shadows, women continue to create pathways to agency as they navigate daily life.

Importantly, conservation anthropologists who work in remote regions are encouraged to work reflexively to shape their research to attend to the historical and layered relationships between humans and the ecosystem on which they depend for cultural, economic, and nutritional sustenance (e.g., Peterson et al., 2010). It is vital to recognize that training in western epistemological approaches to science and knowledge production rooted in a single discipline often cast their own shadows onto the discursive practices that shaped the daily lives of the communities with which we worked (e.g., Chandler & Reid, 2020).

Content, Context, and Consequences of Conservation in Dzanga-Sangha Protected Area

The cascade of choices made to protect wildlife, particularly elephants (*Loxodonta africana*) and other charismatic megafauna (e.g., western lowland gorilla [*Gorilla gorilla gorilla*], forest buffalo [*Syncerus caffer*]), in DSPA was woven intricately into CAR's past, present, and future (Daspit, 2014). Since independence in 1960, CAR has been identified by different publics as a "fractured" or "failed" state, which was consumed cyclically by political upheaval from within and outside of its borders (e.g., Carayannis and Lombard, 2015; Lombard, 2016). CAR's geographic and geopolitical centrality means that the country and its inhabitants have experienced continual ebbs and flows of human migration for hundreds of years, which included Europeans (e.g., French, German, Belgian), internally displaced peoples and environmental refugees, peacekeepers of the United Nations Multidimensional Integrated Stabilization Mission in the Central African Republic (MINUSCA), artisanal and industrial mining concessionaries (e.g., gold, diamonds), and international environmental nongovernmental organizations (Hardin, 2000; Lombard, 2016). Although a complete geopolitical history of the CAR is beyond the scope of this chapter, it is important to acknowledge this history's importance because it shapes the spaces in which conservation, natural resource management, and environmental interventions operate today.

The Wildlife and Conservation Economies of Southwestern CAR

Historical Ebbs and Flows

The regional wildlife and conservation economies of DSPA have ebbed and flowed and so has the impact of change on local livelihoods and women. These changes, particularly the establishment of an integrated conservation and development project (ICDP) (i.e., biodiversity projects with rural development components), have shaped the shadows from which women's work has emerged in contemporary contexts. The interested reader may reference Jost Robinson (2012), Jost Robinson & Remis (2014), or Remis & Jost

Robinson (2020) for deeper insights into the history and impacts of human–animal interactions in DSPA.

The Gazettement of the Dzanga-Sangha Protected Areas Complex

In the 1970s and 1980s, wildlife populations in northern CAR, particularly large mammals such as elephants, experienced extreme hunting pressure, because illegal and militarized networks harvested these species to subsidize geopolitical conflict in the region (Hardin, 2000; Hardin et al., 2014). In the hopes of locating an area within the country where wildlife populations were still intact, researchers funded by the World Wild Fund for Nature (WWF) conducted regional wildlife census work to assess the biodiversity in the southwestern portion of the country (Carroll, 1998). These surveys documented large populations of forest elephants (*Loxodonta cyclotis*), western lowland gorillas (*Gorilla gorilla gorilla*), chimpanzees (*Pan troglodytes*), and bongo (*Tragelaphus euryceros*), which resulted in national and international support for the official establishment of the protected area (Carroll, 1988). The DSPA Complex was gazetted officially in 1990 as a multi-zoned ICDP with support from the CAR government, World Bank, WWF, and other international conservation agencies. DSPA is part of a transboundary protected area known as the Sangha Trinational (STN) where CAR, Cameroon (Lobéké National Park), and Republic of Congo (Nouabalé-Ndoki National Park) meet (Figure 9.1). The "project," as it is known locally, was headquartered in the central town of Bayanga. There were 12 communities within DPSA that were located along a north–south logging road that bisected the complex. DPSA was innovative in its conception and design. It was one of the first protected areas designed as an ICDP aimed at balancing wildlife conservation initiatives with local development needs, and it was the first protected area in CAR that allocated 90% of fees generated from tourism to be put back into the protected area (Carroll, 1992). Although 50% of these funds went into financing the protected area, 40% was allocated to support the local community (Blom, 2001; Carroll, 1998; Hardin, 2000). The way that these funds ultimately were accessed by communities continues to impact the lives of men and women directly who live in the conservation economy, as this chapter discusses below.

DSPA, like many protected areas, is situated in a remote region of the country, at the Upper Sangha River Basin and the extreme southern border with Cameroon and Republic of the Congo. The Sangha River, a tributary of the Congo River, has its own storied history that is rooted deeply in regional politics and economies of resource use. The area has experienced continual migrations of both humans and wildlife that included western lowland gorillas, African forest elephants, bushpigs (*Potamochoerus larvatus*), dark-crowned forest eagles (*Stephanoaetus coronatus*), and blue-breasted kingfisher (*Halcyon malimbica*) (Coquery-Vidrovitch, 1998; Eves et al., 1998; Kretsinger & Zana, 1996).

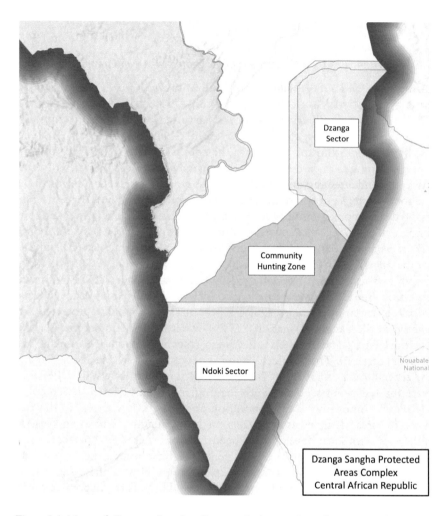

Figure 9.1 Map of Dzanga-Sangha Protected Areas Complex, Central African Republic.

In areas like DSPA that experienced continual waves of migration, it is difficult to define "community," which makes local resource management at the community level challenging (Eves et al., 1998). Before conservation became one of the prevailing regional economies, the communities that lived within the area were immersed in the cyclical presence of extractive industries (e.g., logging, diamond mining, colonial ivory trades) and agriculture (e.g., coffee, rubber plantations) (Blom, 2001; Giles-Vernick, 2002; Hardin, 2000). For the contemporary experiences of women who navigated daily life, legal and illegal wildlife economies within DSPA were connected to these histories. The historical demographics of this area underlay diversity among local communities.

Residents of DSPA belonged to one of two primary categories: Baka and Bilo. Baka belong to a distinct group of Congo River Basin hunter-gatherers; Bilo is a term used regionally to categorize broadly all non-Aka farmers and fishers (Cavalli-Sforza, 1986; Kretsinger & Hardin, 2003). These terms failed to capture the heterogeneity of the indigenous people and local communities. A 2005 census identified 11 ethnic groups and over 30 linguistic groups present in DSPA, which included the Baka, Mpiemu, and Sangha-Sangha. All groups are considered indigenous to this area and are credited with originally founding the town of Bayanga, which was located within DSPA's most southern reaches, at the beginning of the 19th century (Giles-Vernick, 2002; Hardin, 2000; Kamiss, 2006; Kretsinger & Zana, 1996). The majority of migrant groups came to the region to seek work in extractive industries, such as gold and diamonds (Daspit, 2011; Kamiss, 2006). If you walk through the neighborhoods of Bayanga, a well-trained ear can hear shifts in spoken dialects that denote informally the boundaries of each neighborhood, which mark important social boundaries of space. In the past, some spaces were dominated by a single ethnic group. For example, in 2008, Daspit (2011) reported that the majority of market women were of the Gbaya ethnic group—an ethnic group that originated from outside of DSPA that was associated with life in more open, savannah environments. We expected that the representation of indigenous groups, local communities, and women in the DSPA and Bayanga's markets continued to shift, particularly following the 2012 coup d'etat when rebels used the park, its headquarters, and vehicles for non-conservation purposes (Neba & Greer, 2014).

Impacts of Gazettement

The impacts of the innovative model planned for DSPA were muted by the cyclical instability of the region's limited formal economies. Five years prior to DSPA gazettement, the first and longest-running logging company in the area closed its doors. During operations, Slovenia Bois, which was established in 1972, created a network of roads that ultimately connected the DSPA region to surrounding areas and attracted migrants in search of work (Blom et al., 2004). The opening of the roads allowed inhabitants of the region to penetrate further into forested regions, which led to increased legal hunting and extraction of non-timber forest products (NTFPs; Hardin and Remis, 2006). Increased hunting to feed expanding human populations as a result of in-migration to this remote zone marked a turning point in the increasing commercialization of regional wildlife economies. Bayanga residents mostly remembered Slovenia Bois's presence as a time of growth and economic development (Daspit, 2011; Hardin, 2000). At the time that DSPA was established, industrialized logging was, and had been, the prevailing formal economic opportunity in the region (Figure 9.2).

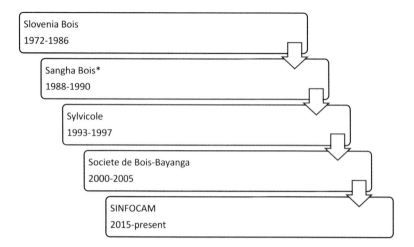

Figure 9.2 Timeline of logging concessions in Dzanga-Sangha Protected Areas Complex. *Indicates establishment of the protected area.

The implementation of the ICDP was intended to generate complementary and enhanced economic opportunities for communities in and adjacent to DSPA, with the intention that logging in the region would become restricted. Logging remained active periodically in Bayanga after DSPA's gazettement, with a succession of company takeovers that had significant impacts, both negative and positive, on the local economy. When each company closed, the families who had relocated to Bayanga and to the DSPA region became unemployed; most stayed in the area and hoped for other economic opportunities. In many instances, informal employment was the prevailing form of economic engagement, and locals participated in farming, fishing, and hunting in the natural resource-rich area.

When Société de Bois Bayanga (SBB) closed in 2005, it was estimated that over half a million USD per year was lost from Bayanga's economy, an amount that was replaced by conservation or tourism (Daspit, 2011). At the time when SBB closed, formal wage labor jobs were available only in DSPA in very small sectors of tourism and research. In 2015, after a period without active logging, the African-based logging company SINFOCAM established its headquarters in Bayanga, which provided much-anticipated formal employment opportunities. The cycle of formal and informal employment opportunities engendered by each logging company's opening and closing brought waves of migrant workers and their families into the protected areas complex, which increased pressure on the natural resources of the region.

These economic interventions only afforded formal work opportunities for men, but created spaces of uncertainty for women. Additional employment opportunities for men altered pressures on women to achieve, new, diverse, and innovative ways to provide for their families. The larger numbers of salaried workers in the area drove business in local formal and informal markets, increased and created new demand for wild game meat and other NTFPs, and created opportunities for women to generate supplemental cash incomes.

The unstable logging economy was coupled with repeated rezoning of forests and changing objectives of conservation organizations at DSPA, which created an economy for local residents that was consistently volatile. The transition from a one-company logging town to a two-company town, with the arrival of conservation economies, led residents to speculate about the future of the region, the "companies," and the livelihoods of local residents (Hardin, 2000). Although the DSPA offered employment to local community members, its employees were primarily men, and jobs were much fewer than those offered by logging companies that occupied Bayanga (Daspit, 2011). Ultimately, the unpredictable economic situation favored and destabilized male labor while increasing stability of women's work. Across the years, women demonstrated the ability to shift rapidly among and between multiple types of work, which included agriculture and commerce, in addition to managing a household.

Separate Income Streams

Generating independent income was critical for women; men and women typically had separate income streams (Kamiss, 2006). Women across DSPA supported their families traditionally through the agricultural production of manioc and commerce derived from, but not limited to, the sale of wild meat and other NTFPs, such as honey, medicine, oil, and silk (Daspit, 2011). Although men, particularly those engaged in the formal economy (e.g., commercial logging, gold mining), generally contributed a lump sum of money to support their families' needs, such as educational expenses, the primary responsibility of providing day-to-day necessities (e.g., food for the household) lay with women. This was especially true for women whose husbands were gone for long periods for work in the forest; these women had to find ways to cover daily household expenses and any unexpected costs that occurred, such as medicine.

Integrated Conservation and Development Projects (ICDPs)

When it was gazetted, DSPA was among a handful of protected areas that was newly established as an ICDP. In concept, ICDPs were to protect biodiversity while simultaneously addressing the needs of local populations who depended on forest and wildlife (e.g., rural development). With this framework, the stimulation of local economic development, which subsequently increased economic stability, was expected to support the protection of wildlife ultimately. As such, development was seen as a means to increase the successes of conservation. In this way, conservation was presented as in the best interests of local people rather than outsiders. Hardin (2000) noted that DSPA was unlike many other ICDPs in that it was underpinned by a commitment to allow indigenous populations to access resources using traditional means in certain sectors, and it allowed for 90% of revenues generated by tourism for the functioning of the project (50%) and for community distribution (40%). The implementation of the original ICDP plan was hampered by the unpredictable and transitory nature of project managers (both expatriate and national); each new wave of personnel brought a unique approach to oversight of DSPA and how to implement policies on resource use by indigenous groups.

In 1991, local Peace Corps volunteers and supporters from the MacArthur Foundation supported development of the Association Communale Yobé-Sangha (ACYS). ACYS was designed to engage a group of local community members who supported conservation, and it used inclusive, participatory models to distribute $5000, developed local trade associations, and generated new revenue streams (Hardin, 2000). ACYS enabled formation of multiple formal and informal associations, which included some that were centered on the efforts of women (e.g., market women selling meat and other food stuffs, restaurateurs, and seamstresses). In total, 21 organizations received funds, which represented all residents (e.g., local/migrant, men/women, all ethnic groups). Importantly, the long-term residents of Bayanga, which included the associations founded by women, received less of the funding from this initiative than more recent migrants (Hardin, 2000). Local political conflicts led to the dissolution of ACYS and the development of alternative mechanisms for distributing revenues, the outcomes and longevity of which were unclear and remained understudied.

Other ICDP initiatives that were designed to be more inclusive of women were met with limited success, which included sustained attention to the development of formal associations and a midwifery training program (Hardin 2000). Opportunities for the inclusion of women were more prominent in the tourism sector. For example, Baka women participated in net-hunts with tourists and provided guided nature walks to show outsiders medicinal plants and other important NTFP. Such an IDCP intervention was aimed at women, but ultimately it was mediated by men because all tourist excursions included a male tour guide. Hardin (2000) illustrated the roles that male tour guides (primarily Bilo) played in arbitrating the potential tips received by Baka during these excursions.

Some women in the DSPA community felt excluded and expressed a haunting sense that the park, which was referred to locally as "the project," was not for them. As one woman put it "[DSPA] gives work to men, but it does not help women" (Jost Robinson, Unpublished raw data). Although some ICDP initiatives attempted to engage with local communities in Bayanga, some women were still on the outside (Daspit, 2011). This may have been because men navigated their own paths autonomously within ICDPs, and women felt that these initiatives were not designed explicitly to include them. Importantly, as women navigated these paths, they often intersected with broader economies of wildlife that followed both legal and illegal trajectories in the shadows of the male-biased structures in a conservation economy.

How Legal and Illegal Wildlife Economies Helped to Shape Life in the Shadows

DSPA, like many other protected areas, monitored and assessed conservation outcomes, which included the functional role of wildlife economies and interventions to support them, such as anti-poaching operations and law enforcement. Ethnographic interviews illuminated the ways that conservation policies and interventions influenced the creation of shadowy spaces where women's work not only supported family needs, but also crossed boundaries between legal and illegal wildlife economies. Importantly, this ethnographic exploration produced unintended insights. For example, some female study participants noted that women were often not monitored or tracked in the same way as their male counterparts. This occurred, in part, because of a lack of transparency in who was responsible for monitoring the sale of wildlife, which took place outside of the integrally protected areas; male game rangers also experienced intrinsic difficulties in policing their "mothers and grandmothers." Interview excerpts that were based on the lived experiences of respondents A and B illustrated the diversity of ecological connections and situational contexts of women who lived in DSPA. The histories and life experiences of women in DSPA were distinctive, and understanding the unique challenges faced by women necessitated understanding the different contexts of their lives. The context of their lives in a conservation economy was what often left women and their husbands wavering between decisions on whether to engage in potentially illegal trades, like snare hunting, or not.

Respondent A's Story

Within the town of Bayanga, houses were clustered closely together at the center, but expanded and spread out with distance like an outstretched hand as one moved north and south along the Sangha River. Many longer-term residents of the town resided in the more central neighborhood; study participants from these central neighborhoods were either born in Bayanga or had lived in the town for most of their lives. One woman in particular, Respondent A, was born in Bayanga after her family relocated from Nola, which was a town north of DSPA. Her father had come to Bayanga after securing formal employment with the Slovenia Bois logging company.

Respondent A reported she rarely had money to spare; when she did, she often paid for a seat on one of the small cars that made the three-hour drive to Nola to visit her remaining family—some aunts, cousins, and her daughter, who was in high school when the interviews were conducted. Bayanga is the center of Respondent A's livelihood. She lived a few houses away from her husband's family and just one neighborhood over from where she grew up.

Her home, which was constructed of older wood planks that sagged because of the humidity, was strategically close to the Sangha River. The river provided crucial water for her household and for processing manioc to sell at the market. Working at her farm kept her away from home for most of the day. Her husband worked at a local tourist lodge that also kept him away from home during the day. Anna's mother and younger sisters were close enough to help manage her household. Despite her husband being employed, Respondent A reported that she felt the weight of much of the day-to-day economic responsibilities for the family. Her husband's monthly paychecks were just enough to cover the daily costs of living, but could not cover any extra expenses. She once commented that once her husband's paycheck was "gone for the month," it was her work that put food in her children's mouths and clothes on their backs. When there were school fees, school uniform costs, or medical costs, it was up to her to secure the funds. Funds usually came from profits from her small farm. If she had extra funds, she would provide them to nearby extended family.

Although Respondent A's manioc farm helped provide staples for sustenance and economic income for her household, the presence of elephants around the farm increased the risk of crop raiding and loss. Indeed, crop raiding by elephants can demolish a farm, and thus a family's entire source of income, in a single night. Respondent A, like other women whose farms had been destroyed by elephants, noted that "the project protects elephants but does not protect the people." She lamented that if her husband was not a DSPA employee, the park would forget that she existed altogether.

Respondent A's story provides a reference for ways of thinking about women who have large, long-standing kin and social networks at the center of the main town in DSPA. Despite having a strong familial network and a husband employed by DSPA, she identified a consistent balance between providing for her household, feeling financially strained, and reporting that the project had forgotten about her. Respondent B's story advanced awareness about local women's lived experiences and challenges faced by migrant women who resided in neighborhoods on the edges of Bayanga.

Respondent B's Story

The overcast September sky threatened rain as Respondent B worked quickly to handwash her family's laundry. Her oldest child brought her a bucket of water from the pump. She had returned from a long morning in her manioc field, approximately 8 km away from her small home on the outskirts of the village, where she has lived for two years. Unlike the center of Bayanga, the outer neighborhoods were quiet; the houses were spread out. Some women gathered together at neighbor's homes to make sweets to sell or to prepare manioc to dry in the sun, but many sat at home without the company of neighbors. Tending to laundry was just one of many tasks for which Respondent B was responsible throughout the day.

Respondent B and her husband had migrated to Bayanga from a small village outside of DSPA when her husband gained employment at SINFO-CAM, the most recent logging company to open in the region. With the rain coming, Respondent B hurried; she still had to go into the forest that day to check the wildlife traps for her husband. On weekends, when her husband was home from his work with the logging company, he took her into the forest to show her how to make and to lay metal wire snares for catching forest animals such as cane rats (*Thryonomys swinderianus*), porcupines (*Hystrix cristata*), and duikers (*Cephalophinae* spp.) (Wire cable snare hunting is illegal in CAR, and the sale of materials used to make snares was prohibited within DSPA.). Nonetheless, snare hunting (a predominantly male activity) was common practice in DSPA. A cheap form of hunting, it required little skill in wildlife tracking and could be either a fallback strategy or a primary occupation for residents (Jost Robinson, 2012). The illegal snare-caught wild meat provided the primary source of protein for Respondent B, her husband, and their three youngest children who lived with them. Although Respondent B noted that she did not regularly sell meat, she stated that if she needed to, she could always sell some of the meat to have money for other things.

Each week, a large truck took Respondent B's husband and other workers into the forest for days to extract timber. When formal work took her husband away from the household, his regular duties were tasked to her without any extended familial support to which she could turn for help. Respondent B expressed appreciation that her husband was employed formally, yet she knew that much of her family's financial obligations, which included school fees and health care, fell to her. Respondent B noted that checking snares was an additional almost-daily responsibility. Snares had to be checked frequently to prevent wastage and to gather necessary protein in time to prepare the family's main meal of the day. Respondent B lamented how checking snares was a very time-intensive, but important responsibility; the week prior, typhoid had kept her in bed and away from her duties.

Lead Author's Story

A major contribution of ethnography to conservation is the ability of the method to enable reflexivity in research (e.g., Foley, 2002). Reflexivity helps to identify assumptions inherent to the research process and researcher positionality; it also enables unique data for triangulating results (Kleine, 1990). The lead author has conducted previous research on coupled relationships between hunters and prey species within forest systems, which included ethnography, interviews, and hunter offtake survey work with male DSPA hunters, 2008–2009 (e.g., Jost Robinson, 2012). Data were coupled with ecological monitoring and diurnal and nocturnal line transect censuses to help assess the population status of mammal species across park sectors (e.g., Jost Robinson et al., 2017). The lead author was the sole female on the research team of 13 and conducted "net drives" with her research team, which attempted to replicate Noss' (1995, 1999) foundational work that demonstrated that accompanying Baka resident hunters in DSPA was a useful method for monitoring species targeted for wild meat. It was only through this research process that the lead author reflected on how the lived experiences of Baka women were being excluded from analysis. Specifically, the lead author noted that a last-minute attempt to reduce high logistical costs of field research trips into the forest resulted in the decision to integrate net surveys with ecological line transect data collection. After the research team returned from the first "integrated" trip, with data, the Lead Author met with a representative group of hunter's wives in Sango. The small group of women requested a daily rate payment for the time she was in the forest with their husbands. They explained to the Lead Author that in taking their husbands into the forests, they had also taken nets into the forest. Net hunting among the Baka people in CAR was a collective practice that required large groups of participants to surround patches of forest with large nets. In taking the nets and their husbands to the forest, the Lead Author had [inadvertently] eliminated an important potential source of protein and an activity for this group of women that generated income. When husbands were away working (e.g., to assist an outside researcher with field research), women often formed all-female hunting parties (Noss & Hewlett, 2001). The lead author's experience illustrated the dynamic autonomy and choice that women made when navigating circumstances and how they were often excluded. In these shadows, women in DSPA renegotiated their situations to adapt to the shifting cultural, economic, and social conditions that stemmed from economic opportunities and conservation interventions.

Inclusive Interventions with Fewer Shadows

This chapter illustrated a diversity of ways that women in DSPA used to shape their daily lives in response to shifting formal economies, which included those generated by ICDPs. In many ways, the case of DSPA in CAR is emblematic of the shadows cast on different portions of different communities at different times. Deeper exploration of these shadows, for example, with a feminist political ecology perspective, created new understanding about the context and the choices people made. Such insight is critically important for designing successful interventions that respond to or help to prevent illegal wildlife economies. For example, on one unusual occasion when (illegally hunted) elephant meat was circulating in the neighborhoods of Bayanga, it was sold and resold quickly as women rushed to try to make a small profit from the commodity. Elephant meat is known to be both illegal and rare, and women talked openly about the situation around the cooking fires and whispered about it along the paths between homes. In this shadowy space, women demonstrated choices to engage in an illicit economy that was fleeting and yet fruitful. It is incumbent upon researchers and policymakers to consider the everyday obstacles and barriers that serve as preconditions to these moments for women in protected areas if sustainable conservation and development is to be successful.

There are practical implications from this conservation anthropology and feminist political ecology-based research. First, there is an opportunity to design more inclusive ICDPs that attend directly to equity. Importantly, equity could (and should!) not be limited to gender inclusivity because it also considers differing values, understandings, and forms of interactions between people and their environments. For example, a lack of attention to multiple forms of inclusivity has limited the success of programs across the world that sought to provide protein alternatives to wild game meat (e.g., Coad et al., 2019). Creating equitable and inclusive ICDPs requires the direct involvement of representatives of the community in every stage (e.g., Romero et al., 2012). In practice, this may manifest as expanded and different opportunities for women, and all community members, to have their voices included during decision-making processes about "alternative solutions" for issues that affect them. Simply including women is insufficient for authentic change because it asks women to move into spaces created by male-focused interventions, which increases pressure on women to occupy multiple spaces at once—a summative burden of negative impact. For example, working in DSPA as a guard can take women away from family responsibilities and increase their reliance on family or social networks for survival, which places new stress on other women in their social networks. Although certain women welcome the opportunity to take on these positions, not all do and, thus effective community-engaged programs can create new and more supportive structures for individuals in these spaces.

DSPA is not starting from zero; existing female-run organizations and associations targeting rural development issues, such as health care, have worked to target these issues. However, these organizations are not acknowledged or supported fully by conservation, development, and government agencies, although sustained support has been provided to men's associations in this area. For example, since 2009, the Bayanga market has seen an increase in the number of female meat sellers (Daspit, 2011). Although the women desired to form a formal association, they have had difficulty achieving the status of a formal association. Formal association status would allow the women to ensure that the organization's members adhered to the regulations of both the association and DSPA. The status would also grant them access to financial support through federal and international aid programs that are available periodically in the area. A regional registered gun hunter's association functions in a similar role, but pertains only to the actual hunting process and not to the sale of meat.

The continued demand for wild game meat, sourced both legally and illegally, is evident by the emergence of a public, evening market for the sale of meat that arrives in the villages around mid-day. The increased opportunities for the public and private sale of wild game meat mean that there are increasing incentives for individuals in DSPA to participate in this aspect of a wildlife economy; yet market women have no formal way to monitor, regulate, and comply with the activities in the formal market. Formal and sustained support for associations could ensure a more sustainable and legal economy of wild meat for local consumption.

In sum, by designing, implementing, and evaluating ICDPs, or other interventions, interventionists can recognize and address more directly the ways engagement can cast opportunity and shadows differently across stakeholders; this leads to unbalanced interventions that create new barriers and sustain old barriers for some individuals. Importantly, the "communities" for which many solutions are designed do not, in practice, exist. The application of anthropological methods, which include ethnographic assessments coupled with feminist political ecology, brings new practical understanding about women's dynamic autonomy and choice in the shifting wildlife economies of CAR.

Literature Cited

Agarwal, B. (1992). The gender and environment debate: Lessons from India. *Feminist studies*, *18*(1), 119–158.

Agarwal, B. (2009). Gender and forest conservation: The impact of women's participation in community forest governance. *Ecological economics*, *68*(11), 2785–2799.

Blom, A. (2001). Ecological and economic impacts of gorilla-based tourism in Dzanga-Sangha, Central African Republic. Master's thesis. Netherlands: Wageningen Universiteit en Researchcentrum.

Blom, A., van Zalinge, R., Mbea, E., & Heitkönig, I.M.A. (2004). Human impact on wildlife populations within a protected central African forest. *African Journal of Ecology*, *42*, 23–31.

Carayannis, T., & Lombard, L. (2015). *Making Sense of the Central African Republic*. London: Zed Books Ltd.

Carroll, R.W. (1988). Relative density, range extension, and conservation potential of the lowland gorilla (*Gorilla gorilla gorilla*) in the Dzanga-Sangha region of southwestern Central African Republic. *Mammalia*, *52*, 309–323.

Carroll, R. W. (1992). The development, protection, and management of the Dzanga-Sangha Dense Forest Special Reserve and the Dzanga-Ndoki National Park in south-western Central African Republic. Dzanga-Sangha Reserve Project and World Wildlife Fund, Washington, D.C., United States.

Cavalli-Sforza, L.L. (1986). *African Pygmies*. New York: Academic Press.

Chandler, D., & Reid, J. (2020). Becoming Indigenous: the 'speculative turn' in anthropology and the (re) colonisation of indigeneity. *Postcolonial Studies*, *23*(4), 485–504.

Coad, L., Fa, J.E., Abernethy, K., Van Vliet, N., Santamaria, C., Wilkie, D.,... & Nasi, R. (2019). *Towards a Sustainable, Participatory and Inclusive Wild Meat Sector*. Indonesia: CIFOR.

Cookson, T.P. (2018). *Unjust Conditions: Women's Work and the Hidden Cost of Cash Transfer Programs*. Oakland: University of California Press.

Coquery-Vidrovitch, C. (1998). The upper Sangha in the time of concession companies. In *Resource Use in the Tri-National Sangha River Region of Equatorial Africa*.

Histories, Knowledge Forms, and Institutions. Edited by H. Eves, R. Hardin, & S. Rupp. Yale School of Forestry and Environmental Studies Bulletin 102, Sangha River Network, 78–90. New Haven, CT: Yale University.

Dahlberg, F. (1981). *Woman the Gatherer.* New Haven, CT: Yale University Press.

Daspit, L.L. (2011). Market women in a central African forest reserve: Engendering wildlife commerce and conservation. Ph.D. Thesis. West Lafayette, IN: Purdue University.

Daspit, L.L. (2014). Losing paradise: War comes to a biodiversity hotspot. Retrieved 1 December, 2020, from https://culanth.org/fieldsights/losing-paradise-war-comes-to-a-biodiversity-hot-spot.

Eves, H.E., Hardin, R., & Rupp, S. (1998). *Resource Use in the Trinational Sangha River Region of Equatorial Africa: Histories, Knowledge Forms and Institutions* (Bulletin Series, YSFES). New Haven, CT: Yale University Press.

Foley, D.E. (2002). Critical ethnography: The reflexive turn. *International Journal of Qualitative Studies in Education, 15*(4), 469–490.

Giles-Vernick, T. (2002). *Cutting the Vines of the Past: Environmental Histories of the Central African Rainforest.* Charlottesville: University of Virginia Press.

Hardin, R. (2000). Translating the forest: Tourism, trophy hunting, and the transformation of forest use in southwestern Central African Republic (CAR). PhD thesis. New Haven, CT: Yale University.

Hardin, R., & Remis, M.J. (2006). Biological and cultural anthropology of a changing tropical forest: A fruitful collaboration across subfields. *American Anthropologist, 108*, 273–285.

Hardin, R.D., Remis, M.J., & Jost Robinson, C.A. (2014). From abundance to acute marginality: Farms, arms, and forests in the Central African Republic: 1988-2014. Retrieved 1 December, 2020, from https://culanth.org/fieldsights/from-abundance-to-acute-marginality-farms-arms-and-forests-in-the-central-african-republi.

Jost, Robinson, C.A. (2012). Unpublished raw data.

Jost Robinson, C.A. (2012). Beyond hunters and hunted: An integrative anthropology of human-wildlife dynamics and resource use in a central African forest. PhD thesis. West Lafayette, IN: Purdue University.

Jost Robinson, C.A., & Remis, M.J. (2014). Entangled realms: Hunters and hunted in the Dzanga-Sangha Dense Forest Reserve (DSPA), Central African Republic. *Anthropological Quarterly, 1*, 613–636.

Jost Robinson, C.A., & Remis, M.J. (2016). Baka women's health and subsistence practices in transitional conservation economies: Variation with age, household size, and food security. *American Journal of Human Biology, 28*, 453–460.

Jost Robinson, C.A., Zollner, P.A., & Kpanou, J.B. (2017). Night and day: Evaluating transect methodologies to monitor duikers in the Dzanga-Sangha Protected Areas, Central African Republic. *African Journal of Ecology, 55*, 222–232.

Gore, M.L., & Kahler, J.S. (2012). Gendered risk perceptions associated with human-wildlife conflict: Implications for participatory conservation. *PLoS One, 7*(3), e32901.

Kamiss, A. (2006). Recensement de la population des villages de la Reserve Speciale de Dzanga Sangha. Report to government of the Central African Republic.

Kleine, M. (1990). Beyond triangulation: Ethnography, writing, and rhetoric. *Journal of Advanced Composition, 10*(1), 117–125.

Kretsinger, A, & Hardin, R.D. (2003). Watershed, weddings and workforces: Migration, sedentarization, and social change among the Baka of southwestern Central African Republic. *African Study Monographs Supp l.2, 8,* 123–141.

Kretsinger, A., & Zana, H. (1996). Cartes des companies foresteiers de la Sangha-Oubangui. Souvenirs de Bayanga 1890-1960, Petite Collection d'Archives et d'Histoire Orales. Bangui, Central African Republic: Imprimerie Saint-Paul.

Leacock, E., Abernethy, V., Bardhan, A., Berndt, C.H., Brown, J.K., Chiñas, B.N.,... & Wadley, S.S. (1978). Women's status in egalitarian society: Implications for social evolution [and comments and reply]. *Current Anthropology, 19*(2), 247–275.

Lombard, L. (2016). *State of rebellion: Violence and intervention in the Central African Republic.* London: Zed Books Ltd.

Lowassa, A., Tadie, D., & Fischer, A. (2012). On the role of women in bushmeat hunting–insights from Tanzania and Ethiopia. *Journal of Rural Studies, 28,* 622–630.

Neba, T.F., & Greer, D. (2014). Update: Conflict in the Central African Republic. *Gorilla Journal, 49,* 15–19.

Noss, A.J. (1995). Duikers, cables, and nets: A cultural ecology of hunting in a central African forest. Ph.D. thesis. Gainesville: University of Florida.

Noss, A.J. (1999). Censusing rainforest game species with communal net hunts. *African Journal of Ecology, 37*(1), 1–11.

Noss, A.J., & Hewlett, B.S. (2001). The contexts of female hunting in Central Africa. *American Anthropologist, 103,* 1024–1040.

O'Shaughnessy, S., & Krogman. N.T. (2011). Gender as contradiction: From dichotomies to diversity in natural resource extraction. *Journal of Rural Studies, 27,* 134–143.

Peterson, M.J., Hall, D.M., Feldpausch-Parker, A.M., & Peterson, T.R. (2010). Obscuring ecosystem function with application of the ecosystem services concept. *Conservation Biology, 24*(1), 113–119.

Remis, M.J., & Jost Robinson, C.A. (2014). Examining short-term nutritional status among Baka foragers in transitional economies. *American Journal of Physical Anthropology, 154,* 365–375.

Remis, M.J., & Jost Robinson, C.A. (2020). Elephants, hunters, and others: Integrating biological anthropology and multispecies ethnography in a conservation zone. *American Anthropologist, 122,* 459–472.

Romero, C., Athayde, S., Collomb, J.G.E., DiGiano, M., Schmink, M., Schramski, S., & Seales, L. (2012). Conservation and development in Latin America and Southern Africa: Setting the stage. *Ecology and Society, 17*(2), 17–40.

10 A Gendered Case File Analysis of Reptile Trafficking and Illegal Keeping in Norway

Ragnhild A. Sollund

The Global Illegal Trade in Reptiles

The international and illegal trade in reptiles, particularly live animals, is a conservation crime and animal abuse issue of increasing scientific significance. The United Nations Office on Drugs and Crime (UNODC) noted that the three largest markets for illegally traded reptiles that appeared in seizure records were reptile skins used for the fashion industry, reptile meat, organs, or venom for food, tonic, or medicine, and live animals for "pets," zoos, or breeding (UNODC, 2020). Reptiles are taken, killed, and traded globally; they are transited using diverse transportation methods through countries such as France, Russian Federation, and Thailand and are commonly destined for markets in China, European Union (EU) countries, Malaysia, Sri Lanka, and the United States (UNODC, 2020; van Uhm, 2016). Illegal reptile trade harms animals, creates vectors for zoonotic disease transmission, and has been associated with other serious crimes, such as money laundering (Wyatt, 2013). Reptile skins, scales, and claws are used in fashion products, such as wristwatch straps, shoes, belts, and purses (Arroyo-Quiroz & Wyatt, 2019). Turtles and tortoises are traded widely as symbols of good fortune, traditional medicine, and food (Ades et al., 2000).

Until 2017, only two countries in the world, Iceland and Norway, banned the trade and keeping of live "exotic" reptiles, which was defined as reptiles that were non-native species. Although native reptiles were illegal to take from the wild and keep due to national wildlife legislation in Norway, it was when Norway joined the Convention on International Trade in Endangered Species of Wild Fauna and Flora (CITES) in 1975 that a ban against non-native reptiles was introduced in a regulation that formed part of the new animal welfare legislation. All keeping, trade, and import of reptiles became illegal; violations normally entailed a fine, and then the animals were euthanized or offered to a zoo. After zoos reached a saturation point in terms of which reptiles they had relative to what they needed, euthanasia became routine (Sollund, 2019).

Norway is an interesting case for the reptile trade for at least two reasons: first, because of the ban that lasted for more than 40 years, a historical record

DOI: 10.4324/9781003121831-13

of violations exists. Since 2017, it has been legal to trade and keep 19 species of non-native reptiles, 15 of which are either listed in CITES Appendix II (12 species) or listed as vulnerable by the International Union for Conservation of Nature (IUCN) (3 species). Second, explorations of wildlife trafficking to Norway, as represented by seizure reports from customs, revealed that nearly 90% of offenders were men (Sollund, 2020). It was after the Food Safety Authority (FSA), which was in charge of animal welfare, argued that the ban entailed animal welfare issues that reptile keeping was legalized somewhat, which disregarded CITES law enforcement (Interview Data 2020).

This chapter discusses the gendered dimensions of reptile trafficking and keeping prior to the lifting of the reptile ban in Norway and the animal abuse inherent in this trade. Although gender-as-culture was known widely and presented a spectrum, wildlife trade data rarely disaggregated beyond male/female sex (Agu & Gore, 2020). As such, the male–female binary is a limited, but first step for analysis, which can also shed further light on the issue of animal harm.

Wildlife and Reptile Trafficking

This analysis defines wildlife trafficking as both the legal and illegal trade in animals that were either free-born or that have been under human control for such a short time that they were in all respects still regarded as wild. Wildlife trafficking usually implies only illegal trade. From the animals' perspective (e.g., Sollund, 2019), the legality or illegality of the trade may be of no significance, and therefore, from a perspective of "harm" under the rubric of green criminology (outlined below), my analysis used the terms "wildlife trafficking" and "illegal trade" interchangeably, unless it was necessary to specify the legality or the illegality of the harm.

Much of the global reptile trade is legal, but a significant proportion is illegal (e.g., UNODC, 2020). There are different ways offenders circumvent rules and laws associated with reptile trade. For example, one study that focused on the role of European markets established that laundering reptiles was widespread (Auliya et al., 2016). CITES trade permits may have been forged, the number of individuals in a shipment may have been wrong (e.g., the document stated there were 36 individuals in a shipment, whereas the correct number was 44) (Sollund, 2019), and documents meant to prove the origin of an animal may likewise have been forged such that wild-caught animals were laundered as locally bred (Nijman & Shepherd, 2015). The reuse of permits was a particular problem in Norway (Interview Data 2013). The problems of fraud and corruption in the animals' countries of origin (e.g., paying off customs inspectors or traffickers who dyed the feathers of birds to pretend they belonged to another species) compounded this (Warchol et al., 2003). Another problem revealed by Auliya et al. (2016) was that >90% of the world's reptiles were not regulated by CITES and European Union Wildlife Trade Regulations (EWTR). This implied that non-CITES species, such as nationally threatened or newly discovered taxa, which commonly represented endemic taxa of charismatic species groups that triggered international demand, were not protected from over-exploitation (Auliya et al., 2016; Marshall et al., 2020). One example of such demand was when a frog with the pattern of the Japanese flag on its back was discovered in Colombia. This novelty increased the desirability of the frog among Japanese collectors (Goyes & Sollund, 2018). Listing a species under CITES also drove up demand because collectors wanted to own an endangered or threatened species (Rivalan et al., 2007).

Reptiles can be "wild-caught" or come from breeding facilities. For example, about 2 million live reptiles are imported annually into the United States, while 9 million are exported. The United States breeds 3–4 million baby "pet" turtles annually, primarily for export (Warwick, 2014). Many turtles from Louisiana are exported worldwide; for example, in Italy, baby turtles from the United States have been observed being sold at a market in Catania, Sicily. A study from 2012 exhibited that >85% of individuals and >90% of shipments from the United States were "wild-caught" (Warwick, 2014). Europe, together with the rest of the EU, was a prime market for wildlife trade, which included the reptile trade (Engler & Parry-Jones, 2007; Sollund & Maher, 2015; van Uhm, 2016; Warwick, 2014). Engler & Parry-Jones (2007) stated that live imports to the EU increased from 225,000 in 1999 to 2.2 million individuals between 2000 and 2006. The reptiles were exported from El Salvador, Togo, and Ghana. The receiving countries were mainly Spain, Germany, and Italy. Reptiles were sold at fairs in Europe, like the huge Hamm fair in Germany, which was known for selling wild-caught animals (Sollund, 2013, 2019). This was despite rules that banned CITES Appendix II species from trade in the EU and in Norway if they were wild-caught.

The main focus of this chapter is on the gendered aspects of keeping reptile pets because these were manifest in case file material from the Norwegian police before the lift of the ban. Was reptile keeping among offenders with long criminal careers gendered? Did such motivation involve more animal welfare concerns, and what were the animal welfare consequences of the lift of the ban? The goal of this chapter was to provide a primarily empirical analysis, with a theoretical framework founded in perspectives of green criminology and gender theory.

Green Criminology Perspectives and Hegemonic Masculinity

Green criminology is a subfield of criminology that is concerned with harms and crimes against the natural environment caused by human action and how such harms affect the natural world and human and nonhuman animals (e.g., Beirne & South, 2007; Brisman & South, 2020; South, 2014). One branch of green criminology is referred to as non-speciesist criminology (Beirne, 1999; Maher & Sollund, 2016; Sollund, 2013, 2015, 2019; White, 2013). This field acknowledges that nonhuman animals suffer harms just like humans because they are also sentient beings and, therefore, can also be victimized (Maher et al., 2017; Sollund, 2017; Taylor & Fitzgerald, 2010; White, 2018).

The reason why this is seldom acknowledged is speciesism: the practice and ideology of discriminating against and abusing individuals simply because they are not human (e.g., Nibert, 2002). Speciesism is analogous to other discrimination based on physical features and appearance, such as racism, sexism, ageism, or disability. Although racism, sexism, and other "-isms" are forbidden in most countries, speciesism is embedded in the fabric of most modern societies (Nibert, 2013).

Consequently, few people question or worry about the abuses and harms that are inflicted upon nonhuman animals that are traded in accordance with CITES regulations, despite the fact that nonhuman animals who are abducted from their habitats, stuffed into boxes and cages, and transported across the world are bound to be harmed (Sollund, 2019; Sollund & Maher, 2015; Warwick, 2014; Wyatt, 2013). This harm may be psychological and/or physical. For example, although wildlife trade is legal in some instances, it may involve abusive treatment of animals. The concept of crime itself offers nothing to characterize the act it involves; beyond its definition as a breach of legislation/regulation, it is a social construct (Christie, 2004). Crime says nothing about how this act affected the victims. The direct victims of the wildlife trade, nonhuman animals, are separated from partners, offspring, parents, and flock; they are mutilated, encaged or snared, or even killed. Therefore, many green criminologists are preoccupied not only with the crimes of wildlife trade, but also the harms involved; present research in green criminology has adopted a harm perspective (e.g., Nurse & Wyatt, 2020; Sollund, 2011, 2013, 2019; van Uhm, 2016; White, 2013; Wyatt, 2013).

Environmental harm is best framed in terms of *justice*, which in turn is based upon notions of human, ecological, and animal rights and broad egalitarian principles (White, 2013). Species justice refers to how harm is constructed in relation to the place of nonhuman animals within their environment and their intrinsic right to not suffer abuse, whether this be one-on-one harm, institutionalized harm, or harm that arises from other human action (White, 2013). In this specific research, one element of this conceptualization is important: one-to-one harm when animals are in captivity. However, it is also central who the offenders are, their gender, and why they engage in such crime. Other research has confirmed that men more often than women are the culprits of wildlife crime and animal abuse (Henry, 2004; Nurse, 2020; Sollund, 2020). This chapter explores the concept of how non-speciesist, green criminology related to gender.

Wildlife crime can be discussed within the framework of gender studies, specifically the concept and phenomenon of hegemonic masculinity that is held in various subcultures. Hegemonic masculinity is one theoretical framework for exploring the dominance, power, or formation thereof of men in particular circumstances (e.g., Donaldson, 1993). The accepted form of hegemonic masculinity requires all men to position themselves in relation to it, and this legitimizes ideologically the global subordination of women to men (Connell & Messerschmidt, 2005). Hegemonic masculinity works through the production of exemplars of masculinity – symbols that have authority and meaning despite that most men and boys do not live up fully to such ideals (Connell & Messerschmidt, 2005).

However, there is not only one normative masculinity, as there is not only one normative femininity; they vary within societies, within different subcultures, and social strata, and hegemonic masculinity is relational (Connell & Messerschmidt, 2005). What one subcultural group may regard as acceptable, admired, or expected may be condemned in another group (Sollund, 2020). Although only a few men actually engage in hegemonic masculinity, it still sets normative standards. Hegemonic masculinity may indicate the most "honorable" way of being a man, but this also often involves violence (e.g. Hong, 2000), and violence can be a means to restore one's lost control, sense of autonomy, and male honor (Sollund, 2000). This standard of male honor varies both between societies and cultures and within subcultures.

Nonetheless, hegemonic masculinity provides a normative framework and guidance for how to "do ideal masculinity." What one regards as ideal depends on one's values and culture, but it is an ideal that one will try to approach to be aligned with one's peers. Shared ideals and rationalization also facilitate or prevent criminality (e.g., Sykes & Matza, 1957). Ideal in this context is regarded as the ideal features of a masculine identity among the offenders whose cases are described below.

For the purpose of this research, it is the masculinity ideals that were outplayed in criminal subcultures that were of specific interest because the primary sources of data were penal case files. Although the ideals of masculinity that were seen therein, such as those related to extrinsic signifiers of "toughness" and aggression, were not valid in society in general and, therefore, merited the term "hegemonic," these signifiers of masculinity were still prevalent (e.g., in gang culture and subcultures that involve animal abuse) (Maher & Pierpoint, 2011; Nurse, 2020). Although traditional ideals of masculinity were often associated with aggression, boldness, and violence, femininity ideals on the other hand were often associated with care and empathy (e.g., Donovan & Adams, 1996; Gilligan, 1983).

A Decade of Data Collection on Wildlife Trade

This chapter is based on data collection and analysis done over a ten-year period on the wildlife trade. The research started in Norway in 2010, which was comprised of semi-structured qualitative interviews with police, customs, the Norwegian Environment Agency, and veterinarians from the FSA. The lead author also collected confiscation reports from the Customs Directorate on seizures of products and live animals listed in the CITES Appendices. There was no consistent coding of CITES crimes within the police statistics in Norway, and when they did exist in Customs, they did not distinguish between animals and plants; there was (to date) no overview of such crimes. With the permission of the Higher Prosecuting Authorities in Norway and the Norwegian Centre for Research Data, roughly 800 penal case files that potentially related to CITES were collected. The police had coded these under various laws, such as animal welfare law, wildlife law, import and export law, and customs law. These cases primarily involved keeping and trafficking in non-native reptiles. The purpose for collecting these archive files was to access all cases in all police districts during 2000–2013; however, the selection of case file material did not cover all relevant cases in the police archives during that time span because this depended on how the police coded the cases.

In 2019, the Research Council of Norway's basic research program, FRIPRO, granted funding for this research for the project *Criminal Justice, Wildlife Conservation and Animal Rights in the Anthropocene* (CRIMEANTHROP). This enabled additional research to repeat and to expand the data collection to see how the lifting of reptile keeping and trade ban affected the enforcement of CITES in Norway. This process is currently underway, but most of the data collection was completed. CRIMEANTHROP enabled further analysis of the 800 case file materials that were accumulated previously; however, given the extensive amount of data, the analyses presented herein are exploratory.

Penal case file archives, which included verdicts, provided important data material for studies in green and conservation criminology. They showed how enforcement agencies and the judicial system assessed the seriousness of such crimes, why the offenders committed them, the efforts put forth in the investigation, and the circumstances under which the offenses were committed. These data complemented interviews where, for example, police investigators elaborated on why cases that were under investigation were dismissed, and these data provided follow-up material for other data, such as customs seizure reports (e.g., Sollund, 2019).

The case files of these offenders were usually very extensive; the folder with the description of the files' contents most often contained a several-page list of offenses, witness testimonies, interrogation reports (interviews the police did with the offenders), photographs of reptiles, and copies of fines and verdicts. Because the majority of these offenders were recidivists, the time span that covered their crimes was usually many years. All these cases came from before the lift of the ban in 2017, so all the reptiles in these cases were kept illegally.

Cases that related to other motivations for wildlife trade were omitted. For example, these omissions included when products or animals were laundered into businesses (e.g., Nijman & Shepherd, 2015; Sollund, 2019), such as watchmakers and the pet trade, traditional Asian medicine (TAM) or traditional Chinese medicine (TCM), trophy-hunting, or the accumulation of adornment objects, such as ivory (e.g., Sollund, 2019). With >800 case files, it was not possible to go into in-depth analysis of all of them here; rather, this chapter presents a general picture of the impression they presented, as represented by a few typical cases.

Through the selection of penal case file material, this part of the study covered only relatively historic features of reptile keeping in criminal subcultures before the lift of the ban, not reptile keeping among law-abiding citizens today. However, this research compared findings from the case file material with findings from in-depth interviews from 2012 that were conducted with five people who were otherwise law abiding, but kept and had trafficked reptiles to Norway illegally before the lift of the ban.

The concepts in green criminology and theories within gender theory were used as a foundation for guiding the interviews with offenders (e.g., concerning the motivations they had for keeping reptiles). Quotations were chosen to represent, for example, different motivations (e.g., Sollund, 2019). In regards to the penal case file material, the research included interrogation reports, photos that elicited the extrinsic value of the animals when offenders posed with them and descriptions of the scene (i.e., how reptiles were kept), and verdicts that analyzed motivations and social backgrounds of offenders (i.e., subculture affiliation).

Several aspects of the case file material have been analyzed. The most important aspect was whether or not the file involved wildlife trafficking and animal abuse and how these related to masculinity ideals. For example, an offender with a long criminal record that involved drug crimes and violence was generally regarded as unfit to keep animals, especially when this was supported by photos and descriptions in the case file material that revealed unacceptable conditions.

Reptile Keeping and Trafficking in Norway: Who Were the Offenders?

The police most often apprehended the reptile keepers who were the focus of this chapter because they had been involved in other forms of crime. The police may have performed a house search because they suspected the offenders were involved in, for example, theft, bodily harm, or drug crimes. When the police came to these offenders' houses, they typically found drugs, weapons, and a glass box that contained one or more reptiles. About 90% of offenders in the case files were male, and females accounted for 5–10% of offenses.

As an example of such cases, which are referred to as combination cases, is one from 2008. The male offender had a large repertoire of criminal activity: he was charged with breach of weapon legislation, fraud, receiving and selling stolen goods, car theft, burglary, insurance fraud, several incidents of drug crimes and, finally, having smuggled a bearded dragon (*Pogona vitticeps*) to Norway from Sweden. The reptile was kept in a glass box at the offender's house when the police came to search it. The offender admitted in the police interview that he used amphetamines and smoked cannabis regularly.

According to the police file, the FSA would try to give the animal to a zoo; if unsuccessful, it would be killed.

In another case, the offender was convicted of drug crimes that involved cocaine and illegal liquor production, and for having kept a three-meter-long boa snake (*Boa boa*) for a friend. The police further confiscated dynamite, steroids, illegal liquor, and NOK 10,000 (ca. USD $1,085) in cash. The snake was euthanized. In yet another case, a man was convicted for car theft, theft from cars, and many drug crimes. When the police came to the offender's house, they found a tiger python snake (*Python molurus*) on the offender's sofa in the living room. The snake was about two years old and 180 cm long; it was confiscated and euthanized. In another case, a man was convicted for drug crimes, for the possession of 307 ecstasy pills, 7 g of amphetamine, 5 g of hashish, and other drugs, and for having sold 150 ecstasy pills. He was also involved in car theft and had bought stolen tools. Finally, he had in his possession a tiger python, about 180 cm long. He had been convicted previously of drug crimes. The file did not mention the fate of this snake.

Yet another case involved a man who was arrested for drug crimes, which included amphetamine, cannabis, and undefined pills and for keeping stolen goods and illegal ammunition. He had six lizards and snakes in two glass boxes. These included bearded dragons and boa constrictors that he had trafficked from Sweden. They were later euthanized. It was only with very rare exceptions that confiscated reptiles in Norway were not euthanized. There were no rescue facilities for seized wildlife in Norway, except for zoos that declined offers to rehome confiscated animals if the species was represented in their collection already.

These cases illustrated how wildlife trafficking entailed animal abuse and breach of species justice of which the reptiles were victims. However, in the eyes of the police, such crimes were just one in a long list of offenses that were regarded as more serious, particularly the possession of illegal weapons, ammunition, violence toward humans, and drugs. To possess illegal weapons or to use steroids were signifiers of adherence to masculinity ideals in which aggression and violence were a part (e.g., Hong, 2000; Donaldson, 1993).

One last example of a male offender in the case file material had many features similar to the others, although it also differed from the rest because of the number of animals and other confounding factors. This man was convicted for a long list of drug crimes and for illegal possession of many different weapons and ammunition. When the police searched his house, they found three bearded dragons, one red-eared terrapin (*Trachemys scripta elegans*), six yellow-eared terrapins (*Trachemys scripta scripta*), two albino boa constrictors, one rainbow boa (*Epicrates cenchria*), and one python (*Pythonidae*). All the reptiles were seized and then euthanized by FSA veterinarians. The offender also had in his flat a desert rat (*Dipodomys deserti*), a tarantula (*Theraphosidae*), a nymph parakeet (*Nymphicus hollandicus*), two Fisher parakeets (*Agapornis fischeri*), and a large number of "aquarium fish," according to the file. These animals were left in the flat unattended by police because they were legal, even though the owner was then in custody and, therefore, unable to care for them.

In the previous cases listed, the reptiles were confiscated and euthanized, and according to the police files, the offenders did not protest. In sum: the police came to perform a house search of a known criminal who was connected to a large number of offenses or to apprehend the man. They found a large amount of illegal drugs or weapons and a number of reptiles. They then called veterinarians from the FSA, who usually took the animals and euthanized them, based on advice from the Norwegian Environment Agency. When questioned about the reptiles, the offenders often stated that they obtained the animals "secondhand;" most often the reptiles were snakes, particularly boa constrictors.

In the last example, the offender claimed he had persuaded another man known to pressure other people for money with violence on commission (i.e., locally known as a torpedo) to sell him the terrapins. According to his own testimony, he had received the lizards when he was heavily intoxicated. He claimed to have done the animals a "favor." Despite this, the authorities killed all 16 reptiles. Many of the reptiles in the case file material that were confiscated and killed are now legal. Therefore, in similar cases today, they would most likely survive a police search, but whether that would be to their benefit would depend on their fate in the hands of the offender.

So why did these offenders who were involved in weapons crimes, drug crimes, and other crimes keep these animals? Except for the last offender, who claimed a sense of responsibility toward the animals, the great majority of these offenders exhibited no affection for the reptiles they kept. As it was exemplified in the files, the offenders willingly accepted that the reptiles were confiscated and euthanized, and the description of how the animals were kept demonstrated that there was little interest in their welfare. Snakes were more often masculinity tokens than, for example, a small anole or tortoise, and snakes comprised the majority of the reptiles in these cases.

Relationship of Female Offenders with Their Reptiles

There were also women in the case file material, albeit only 5–10%. One of these women received a visit from the police because they suspected her of illegal drug possession. She had a snake she called Lisa, species unknown. The woman was very upset when the police seized the reptile, especially because Lisa would be euthanized. She was evidently fond of her. She said: "Please do not take her away from me, she is my baby." She did not accept that she should be killed. In most of the cases that involved women and reptiles, they trafficked a reptile to Norway and had been stopped at the border. For example, in a few cases, a terrapin baby was bought in a market abroad (e.g., Turkey) and hidden in a pocket during the flight. In contrast to the large majority of the cases – such as the combination cases described previously – most women charged for illegal possession or keeping of reptiles had no other charges against them.

Another woman in the case file material, who was also interviewed, received a police visit because a neighbor suspected she was growing marijuana due to the UV lighting seen from outside. The police found no plants, but many reptiles of different species, which included iguanas (*Iguana iguana*), boas, and bearded dragons: she had about 20 at the time. She said she took on the role of "mom," was a rescue home for abandoned reptiles, and that she had a reputation as such. She also committed a crime when she brought a couple of reptiles from abroad, and she had bought others illegally in Norway prior to the lift of the ban. When she was interviewed, she stated she had kept reptiles for 20 years. She said:

> So many animals have been by. I have become a kind of dumping site for reptiles, because I am so frequently on the [reptile internet] forums. I have spent a lot of time teaching people about how to keep these animals. I think behavior is very interesting, whether of animals or humans. So, as people get tired of animals, they dump them at my door. They knew where to go, and I couldn't say no. If the alternative is euthanasia or that someone else takes them, then the answer is a given.
>
> (Sollund, 2019, p. 126)

She also said that she used to have a boa, her first, which she used to take outside, and when they were out, the snake would wrap herself around her waist. She described how the snake would crawl up to her face when she was sitting on the sofa, all the way up to her glasses, which the snake would then ruin (e.g., from knocking them down or bending them). She had wrecked many pairs. This woman's relationship to the reptiles indicated nearness, responsibility, and care.

Another woman who also figured both in the case file material and as an interviewee similarly expressed care for her reptiles, as exemplified when she said that she loved to buy special treats for them and to watch them eat (Sollund, 2019). So, while few, the women in the data expressed concern for their reptiles. The relationship these women had to their reptiles was in stark contrast to the hegemonic masculinity ideals that surged from the data material when male offenders engaged in "masculinity" crimes and when their motives for keeping reptiles became evident. Although "toughness and aggression" were characteristic of male-dominant subcultures, caring was more of a feminine trait (e.g., Donovan, 1996; Gilligan, 1982). This type of "toughness" and aggression may have been direct determinants of abuse and species injustice, and their motivation for keeping reptiles may have been as well.

The Meaning of a Reptile: An Extrinsic Token of Masculinity or Someone to Care For?

In some cases that involved male offenders, the reptiles, particularly snakes, seemed to have been used as extrinsic tokens or as fetishes (Collard, 2020). Similar to the affinity that many of these offenders seemed to have for weapons, which empowered them and created fear in other humans, they may have used big snakes for the same purpose. In the case files, there appeared, for example, photographs from YouTube videos the offenders had posted of themselves, with the snake around their head and shoulders, or posing with the animals in other ways. The combination of tattoos, large muscles, and a big snake were meant to be tokens of "toughness" and "masculinity." The pictures in the files proved the offenders' engagement with a presumed "dangerous" species, which were regarded as dangerous in large part *because* they were illegal; that may itself add to their "toughness status" in criminal subcultures. Nonetheless, through this posing, one may suspect they attempted to enhance their personal image, which was akin to narcissistic trophy photos or to keeping "dangerous dogs" (Maher & Pierpoint, 2011). With their prehistoric appearance, reptiles often elicit little empathy and are often alienated and objects of phobias. As observed by Öhman & Mineka (2003), fear for reptiles is likely a core mammalian heritage. From this perspective, snakes and other reptiles have a special psychological significance for humans.

Offenders also displayed "toughness" by posting videos of how they fed snakes with live animals. To use live feed was illegal in Norway, and it was a breach of the Animal Welfare Act. One such case ended in court. The verdict was handed down in Follo District Court, where the offender received a 24-day unconditional prison sentence for feeding his 2-m-long boa live animals on several occasions. He had filmed himself feeding the snake a live rabbit and a live hamster. His posting of these films indicated that he was proud of this. The male offenders were largely disinterested in the reptiles' fate when they were confiscated and euthanized, and they seemed to have accepted it quite easily. Meanwhile, as was demonstrated by the case files, women mentioned motivations for keeping reptiles that related much more to care and mutuality. Some case files noted women expressed grief when the animals were taken from them to be euthanized and sometimes tried to stop euthanasia, for example, in suggesting facilities where they could be rehomed; on occasion, these women even arranged for housing. Although the cases that involved women's illegal reptile keeping prevented any generalization, it is worth commenting further on these features of their relationship to the reptiles.

First, as women's role has traditionally been that of caregiver, when one of the women accepted a role as a dumping site for other peoples' abandoned reptiles, she fulfilled the expectations directed at women (Beauvoir, 2011; Donovan & Adams, 1996; Gilligan, 1982). Many find it "natural" to expect that women would be nurturers and caregivers, even to reptiles, rather than to use these animals instrumentally, as many of the males did when they used them to display status as "tough men" (Fowler & Geers, 2017). The female reptile owners may have fulfilled expectations directed at them as women, although reptile species varied substantially, and some were perceived as having more charisma than others. The interviewees relayed how they felt forced to socialize only with other reptile owners because of the criminalization of reptile keeping, which made them constantly alert to the risk of being revealed by others for this offense (Sollund, 2013). They spoke of living behind closed doors, never opening the door for strangers, and never inviting in new friends for fear they would report them (Sollund, 2013). During the years of the ban, they were forced to form subcultures.

It may be assumed that the male offenders in the case file material also socialized privately mostly with other people within their own group, like most other people do. This suggested that they belonged to a subculture that valued, or at least accepted, the keeping of illegal reptiles or even used them as tokens. Therefore, one can assume that when male offenders exhibited their reptiles in, for example, a coffee table that contained a hole where the snake was kept under a glass screen, as in one case, this was something for which they received social support among their peers. If their peers morally reproached them, they were less likely to keep the snake under such deplorable conditions. The case file material suggested that these male offenders generally received social support in their subculture for the ways in which they used reptiles to enhance their status.

However, this may also be a simplification and valid only in the subcultures to which these men belonged and was unlikely to be representative of all male reptile keepers. Another man I interviewed, whom I have no reason to suspect of engaging in other crimes, also expressed a great deal of care for his reptiles. He enjoyed their company, and he said that he also had a moray eel (*Muraenidae*) that he used to coddle. In the last example of the male cases referred to above, in which the offender claimed to have taken animals as a "favor," he did the same as the woman who took in abandoned reptiles. Consequently, he may have had similar motivations as the "reptile mom," which was to save the animals from a worse fate in the hands of his peers who did not care for them.

According to Statistics Norway, in 2018, far more men than women were involved in drug crimes: 22,653 men compared with 3,297 women. This was also true concerning crime in general: 122,070 men compared with 20,092 women (Statistisk Sentralbyrå [Statistics Norway, SSB], 2018). As mentioned, the great majority of cases in my material concerned male drug offenders who also had a reptile, and who were usually involved in other crimes. Perhaps, however, the police detected them more easily because their crimes were revealed easily, as when they boasted about their reptiles. When they were male drug offenders who also engaged in other crimes, the police became aware of them more easily. The fact that men were overrepresented so heavily in this crime sample indicates a control effect caused by men being highly visible in their crimes and their large contribution to crimes more generally. A woman may have been less open about her crimes and to have committed crimes that were hidden more easily; this might be particularly true of those women who were connected to her reptile keeping because her relationship to the reptile was not instrumental, but one of care. Because no study has been done on this in Norway, it is impossible to say how many women kept reptiles compared with men, before and after the ban.

It *is* possible that reptile keeping per se was not gendered, whether or not reptile keeping was legal or illegal. Perhaps, reptiles held a similar attraction for men and women under other circumstances. For example, although male offenders may have felt a specific attraction to reptiles, particularly snakes, because they enhanced their own "toughness factor" even further because the reptiles were illegal, women generally may have been deterred from buying and keeping reptiles precisely for the same reason, similar to law-abiding men. However, a study by Bride (1998) demonstrated that, among youths, boys kept reptiles more often than girls.

Animal Harm

Some scholars and practitioners have advocated that people who were involved heavily in drug crimes and other forms of crime, like the typical offender in the case file material, should not keep animals. As exemplified by the man who kept a snake in a table, the rationale was based on notions of animal harm and the perception that such offenders did not take the necessary steps to ensure their animals' welfare. The pictures of the reptiles that were often included in the case files documented inadequate terrariums. Animals were often kept in small, bare glass boxes with no form of decoration to provide mental and physical stimulation or a place to hide. There was no appropriate heating or lighting. Animals can suffer direct abuse, but also suffer from neglect (Agnew, 1998; Beirne, 1999). Even under relatively "good" conditions, research suggested that captive reptiles suffered from being kept in terrariums, and the majority of reptile keepers had inadequate knowledge of these species' intelligence, sentience, and needs (Collard, 2020; Warwick, 2014). Herbig (2010) compared captive reptiles to "cut off flowers," which implied short lifespans. It has been established that reptiles suffer in captivity; for example, a study from the United Kingdom showed that four out of five reptiles died after a year in captivity, and generally, reptiles were usually kept under unsatisfactory conditions, which caused premature death (Warwick, 2014).

It is unlikely to be in the reptiles' interest to be used as fetishist extrinsic tokens of "toughness" to strengthen the offenders' personal identities and status. Being stored away in boxes in basements or caged in a closet, as the case file material also indicated, were far from the conditions such animals had in the wild and lacked what was necessary to satisfy their physical and mental needs. When the case file material showed that many of the offenders were also charged with violent offences, there was reason to fear that reptiles in their custody were also subject to abuse. This is not an assumption based on the "graduation hypothesis" that there is a "development" from animal cruelty to interpersonal violence, but is based on the fact that someone prone to use violence against a human is likely to use violence against animals (Agnew,

1998; Ascione et al., 2007; Flynn 2011). Henry's (2004) study on animal abuse and delinquency showed youths who self-reported that they engaged in animal abuse were also involved in other crimes. This study also demonstrated that the men in the study reported having participated in animal cruelty far more than the women studied (27 men compared with three women, 35.1% compared with 3.3%). This suggested there was more reason to be concerned for animal welfare when men were owners than when women were owners and, particularly, when these men were involved in other forms of crimes.

Estimates based on sale of reptile equipment and seizures revealed that prior to the lift of the ban, there were more than 100,000 illegal reptiles in Norway, and the FSA claimed it would be easier for reptile owners to seek veterinary care if reptiles were legalized (Sunde, 2011). On the other hand, reptile owner interviewees said that this was never a problem. Contrary to what was suggested by the FSA, there were many veterinarians who had the necessary skills to treat reptiles and would receive them despite the ban.

Instead of improving the welfare conditions for exotic reptiles in Norway, the lifting of the ban may have caused them to deteriorate because reptiles have become so easily accessible. They can now be purchased in "pet" stores and on the internet through the website Finn.no, where they are sold alongside other typical 'secondhand animals,' such as parrots and rabbits. The only requirement for purchasing an animal in Norway is that one must be older than 16 years of age. However, no one examines the purchases made on Finn.no. The zoo shops may give buyers guidance, but when people sell reptiles on the internet secondhand or after breeding them, such requirements are not necessarily fulfilled. Also, people tire of their reptiles, which may have negative outcomes as well. Greater accessibility of "pet reptiles," just as is the case for other companion animals, entails the risk of abuse for more individuals (Arluke & Irvin, 2017; Sollund, 2011). There is little to suggest that the lifting of the ban resulted in less animal abuse. On the other hand, criminal subcultures may find that with the lifting of the ban, reptiles did not enhance the social status of men who were connected to hegemonic masculinity, such as "toughness", because anyone older than 16 years of age could purchase a reptile in a shop. The animals are now presented as harmless through legalization, which breaches the extrinsic connotations of danger connected to them when they were illegal that may have motivated the offenders to obtain them.

Conclusion

This chapter has focused on the illegal reptile trade and motivations that people have for keeping illegal reptiles, with a special regard to animal abuse, species justice, and gendered ideals among offenders. Based on penal case file material, the majority of offenders who kept reptiles illegally were men who were often involved in other crimes. Whether other law-abiding male reptile keepers also used reptiles as tokens of toughness and masculinity was not clarified by this research. However, based on the case file material, reptile keeping is gendered, at least in criminal subcultures. Case files showed male offenders involved in multiple crimes used reptiles to enhance their social status. On the contrary, the few women who appeared in these data did not use the reptiles in the same instrumental way; rather, they had a relationship to the reptiles that seemed more connected to traditional expectations of femininity, such as care. More studies are needed to assess what kind of relationship men and women who keep reptiles have to their animals and how those animals are treated.

Literature Cited

Ades, G., Banks, C. B., Buhlmann, K. A., Chan, B., Chang, H. C., Chen, T. H.,… & Shi, H. T. (2000). Turtle trade in northeast Asia: regional summary (China, Hong Kong and Taiwan). In *Asian Turtle Trade-Proceedings of a Workshop on Conservation and Trade of Freshwater Turtles and Tortoises in Asia, Phnom Penh, Cambodia, 1–4 December 1999*, 52–54.

Agnew, R. (1998). The causes of animal abuse: A social-psychological analysis. *Theoretical Criminology*, *2*, 177–209.

Agu, H. U., & Gore, M. L. (2020). Women in wildlife trafficking in Africa: A synthesis of literature. *Global Ecology and Conservation*, *23*, e01166.

Arluke, A., & Irvine, L. (2017). Physical cruelty of companion animals. In: Maher J., Pierpoint H., Beirne P. (eds) *The Palgrave International Handbook of Animal Abuse Studies*: 39–57. London: Palgrave Macmillan. https://doi.org/10.1057/978-1-137-43183-7_3

Arroyo-Quiroz, I., & Wyatt, T. (2019). Wildlife trafficking between the European Union and Mexico. *International Journal for Crime, Justice and Social Democracy*, *8*, 23.

Ascione, F. R., Weber, C. V., Thompson, T. M., Heath, J., Maruyama, M., & Hayashi, K. (2007). Battered pets and domestic violence: Animal abuse reported by women experiencing intimate violence and by nonabused women. *Violence against women*, *13*(4), 354–373.

Auliya, M., Altherr, S., Ariano-Sanchez, D., Baard, E. H., Brown, C., Brown, R. M.,… & Hintzmann, J. (2016). Trade in live reptiles, its impact on wild populations, and the role of the European market. *Biological Conservation*, *204*, 103–119.

Beauvoir, S. (2011). *The Second Sex*. New York: Vintage Books.

Beirne, P. (1999). For a nonspeciesist criminology: Animal abuse as an object of study. *Criminology*, *37*, 117–148.

Beirne, P., & South, N. (eds.). (2013). *Issues in Green Criminology*. Devon: Willan.

Bride, I. (1998). Herpetofauna pet-keeping by secondary school students: Causes for concern. *Society & Animals*, *6*, 31–46.

Brisman, A., & South, N. (2020). *Routledge International Handbook of Green Criminology*. United Kingdom: Routledge.

Christie, N. (2004). Hvorfor det er ufruktbart å ta utgangspunkt i at kriminalitet er en sosial realitet. *Nordisk tidsskrift for kriminalvidenskab*, *91*(4), 404–407.

Collard, R.-C. (2020). *Animal Traffic*. Durham and London: Duke University Press.

Connell, R., & Messerschmidt, J. (2005). Hegemonic masculinity: Rethinking the concept. *Gender & Society*, *19*, 829–859.

Donaldson, M. (1993). What is hegemonic masculinity?. *Theory and society*, *22*(5), 643–657.

Donovan, J. (1996). Attention to suffering: A feminist caring ethic for the treatment of animals. *Journal of Social Philosophy*, 27(1), 81–102.

Donovan, J., & Adams, C. J. (1996). *Beyond Animal Rights: A Feminist Caring Ethic for the Treatment of Animals*. New York: Continuum International Publishing Group.

Engler, M., & Parry-Jones, R. (2007). *Opportunity or Threat: The Role of the European Union in Global Wildlife Trade*. Brussels: TRAFFIC.

Flynn, C. P. (2011). Examining the links between animal abuse and human violence. *Crime, Law and Social Change*, 55, 453–468.

Fowler, S. L., & Geers, A. L. (2017). Does trait masculinity relate to expressing toughness? The effects of masculinity threat and self-affirmation in college men. *Psychology of Men & Masculinities*, 18, 176.

Gilligan, C. (1982). *In a Different Voice*. Cambridge: Harvard University Press.

Goyes, D. R., & Sollund, R. (2018). Animal abuse, biotechnology and species justice. *Theoretical Criminology*, 22, 363–383.

Henry, B. C. (2004). The relationship between animal cruelty, delinquency, and attitudes toward the treatment of nonhuman animals. *Society & Animals*, 12, 185–207.

Herbig, J. (2010). The global reptile trade as a form of conservation crime: A South African criminological investigation. In *Global Environmental Harm: Criminological Perspectives*, ed. R. White: 110–132. London: Wilan Publishing.

Hong, L. (2000). Toward a transformed approach to prevention: Breaking the link between masculinity and violence. *Journal of American College Health*, 48, 269–279.

Maher, J., & Pierpoint, H. (2011). Friends, status symbols and weapons: The use of dogs by youth groups and youth gangs. *Crime, Law and Social Change*, 55, 405–420.

Maher, J., & Sollund, R. (2016). Law enforcement of the wildlife trafficking: a comparative strengths, weaknesses, opportunities and threats analysis of the UK and Norway. *Journal of Trafficking, Organized Crime and Security*, 2, 82–99.

Maher, J., Pierpoint, H., & Beirne, P. (eds). (2017). *The Palgrave International Handbook of Animal Abuse Studies*. London: Palgrave.

Marshall, B. M., Strine, C., & Hughes, A. C. (2020). Thousands of reptile species threatened by under-regulated global trade. *Nature Communications*, 11, 1–12.

Nibert, D. (2002). *Animal Rights/Human Rights: Entanglements of Oppression and Liberation*. Plymouth, United Kingdom: Rowman & Littlefield.

Nibert, D. (2013). *Animal Oppression and Human Violence: Domesecration, Capitalism, and Global Conflict*. New York: Columbia University Press.

Nijman, V., & Shepherd, C. R. (2015). *Adding Up the Numbers*: an investigation into commercial breeding of Tokay Geckos in Indonesia. TRAFFIC. Selangor, Malaysia: Petaling Jaya.

Nurse A. (2020). Masculinities and animal harm. *Men and Masculinities*, 23, 908–926. doi:10.1177/1097184X20965458.

Nurse, A., & Wyatt, T. (2020). *Wildlife Criminology*. England: Bristol University Press.

Öhman, A., & Mineka, S. (2003). The malicious serpent: Snakes as a prototypical stimulus for an evolved module of fear. *Current Directions in Psychological Science*, 12, 5–9.

Rivalan, P., Delmas, V., Angulo, E., Bull, L. S., Hall, R. J., Courchamp, F.,... & Leader-Williams, N. (2007). Can bans stimulate wildlife trade? *Nature*, 447, 529–530.

Sollund, R. (2000). På bunnen av hotellet. *Søkelys på arbeidsmarkedet*, 2(17), 245–253.

Sollund, R. (2011). Expressions of speciesism: The effects of keeping companion animals on animal abuse, animal trafficking and species decline. *Crime, law and social change*, 55(5), 437–451.

Sollund, R. (2013). Animal trafficking and trade: Abuse and species injustice. In *Emerging Issues in Green Criminology*, eds. S. R. Walter, D. Westerhuis, & T. Wyatt: 72–92. London: Palgrave Macmillan.

Sollund, R. A. (2015). The illegal wildlife trade from a Norwegian outlook: Tendencies in practices and law enforcement. In *Green Harms and Crimes*, ed. R. Sollund: 147–169. London: Palgrave Macmillan.

Sollund, R. (2017). The animal other: Legal and illegal theriocide. In *Greening Criminology in the 21st Century*, eds. M. Hall, T. Wyatt, & N. South: 79–99. New York: Routledge.

Sollund, R. A. (2019). *The Crimes of Wildlife Trafficking: Issues of Justice, Legality and Morality*. London: Routledge.

Sollund, R. (2020). The victimisation of women, children and non-human species through trafficking and trade: Crimes understood through an ecofeminist perspective. In *Routledge international handbook of green criminology*. Routledge: pp. 512–528.

Sollund, R., & Maher, J. (2015). The illegal wildlife trade: A case study report on the illegal wildlife trade in the United Kingdom, Norway, Colombia and Brazil. *A Study Compiled as Part of the EFFACE Project*. https://efface.eu/sites/default/files/EFFACE_Illegal%20Wildlife%20Trade_revised.pdf. Accessed February 11, 2017.

South, N. (2014). Green criminology: Reflections, connections, horizons. *International Journal for Crime, Justice and Social Democracy*, 3(2), 5–20.

Statistics Norway. (2018). https://www.ssb.no/sosiale-forhold-og-kriminalitet/statistikker/lovbrudde/aar. Accessed on October 23, 2020.

Sunde, S. (2011). Ett av tusen ulovlige reptiler i Norge blir beslaglagt? [One in a thousand reptiles confiscated?] Miljøkrim. Tidsskriftet for miljøkriminalitet. https://www.okokrim.no/ett-av-tusen-ulovlige-reptiler-i-norge-blir-beslaglagt.416812.no.html. Accessed on October 23, 2020.

Sykes, G. M., & Matza, D. (1957). Techniques of neutralization: A theory of delinquency. *American Sociological Review*, 22, 664–670.

Taylor, N., & Fitzgerald, A. (2018). Understanding animal (ab)use: Green criminological contributions, missed opportunities and a way forward. *Theoretical Criminology*, 22, 402–425.

TRAFFIC. (2012). Captive-bred or wild-taken? Examples of possible illegal trade in wild animals through fraudulent claims of captive-breeding. TRAFFIC/WWF.

United Nations Office on Drugs and Crime. (2016). World wildlife crime report 2016: trafficking in protected species. United Nations Office on Drugs and Crime. Vienna, Austria

van Uhm, D. P. (2016). *The Illegal Wildlife Trade: Inside the World of Poachers, Smugglers and Traders* (Vol. 15). Switzerland: Springer.

Warchol, G., Zupan, L., & Clarke W. (2003). Transnational criminality: An analysis of the illegal wildlife market in southern Africa. *International Criminal Justice Review*, 13, 1–26.

Warwick, C. (2014). The morality of the reptile 'pet' trade. *Journal of Animal Ethics*, 4, 74–94. doi:10.5406/janimalethics.4.1.0074.

White, R. (2013). *Environmental Harm: An Eco-Justice Perspective*. Bristol, United Kingdom: Policy Press.

White, R. (2018). Green victimology and non-human victims. *International Review of Victimology, 24*, 239–255.

Wyatt, T. (2013). *Wildlife Trafficking: A Deconstruction of the Crime, the Victims, and the Offenders*. Hampshire, England: Springer.

Box 4

Voices from the Field

Beneficiaries & Persons Harmed

Helen U. Agu and Meredith L. Gore

It is possible for women to derive direct and indirect benefits from wildlife trafficking, and all may accrue in different spaces and places. Income is frequently mentioned as a benefit associated with the crime; however, cultural expression, employment, empowerment, health security, prestige, and recreational benefits may also accrue. Different benefits may amass over time, and the value of these benefits may be weighed against costs or harms. For example, certain wildlife products are used as high value art objects. Such products make excellent gifts, especially for the upwardly mobile. Where these products are under strict international control, their possession may also serve to demonstrate the political connections of the owner. In other words, they convey prestige precisely because attaining them legally is difficult. Person's harmed are individuals victimized or made vulnerable by wildlife trafficking. Beyond immediate family members, it is possible that wildlife trafficking harms orphans, refugees, widows, or un(der)employed people. Harms may be physical, social, emotional, spiritual, or economic, to name a few. When we asked experts their perceptions of how women in wildlife trafficking benefit (study design, methods, and analysis discussed in Agu et al.[1,2] economic examples dominated, as illustrated by these voices from the field:

> Think the benefit is very clear, is that the money aspect of it…. the monetary part of activity is their benefit.
>
> (Male 1.3.2)

> So to me, I think women actually haven't got seriously in smuggling and participating in organized crime anywhere except with the first suspects. People don't expect them to be doing this. So maybe that cost is quite low. So, the benefits are high enough to keep them in the business and it will be very, very, very interesting to study these dynamics.
>
> (Female 1.3.1)

> A lot of as we know, a lot of African communities do not view wildlife as a touristic product because at the end of the day, they're trying to survive,

DOI: 10.4324/9781003121831-14

a lot of them have very low incomes and the benefits of it as If you're able to successfully work in this, you can get quite a lot of money for little quite relatively little effort.

(Female 1.3.2)

And also, I believe there is a tendency for medicinal or medicinal providers to be women in many cultures. And so, in fact, pangolins scales are being used for traditional medicine, I suspect that women are players and maybe it is a benefit for them, because it's a job.

(Male 1.4.3)

Like in any economy, anyone involved in the choppy waters of smuggling of wildlife products or live animals, they get their share. And usually this income is greater than the actual income they make. And that's an incentive for people to out and do the illegal activity and to the best of my knowledge, women are in the middle.

(Male 1.5.6)

The precise motivations and intentions of women were not always clear to experts, but they positioned benefits from wildlife trafficking within women's lived experiences, as some voices from the field help exemplify:

Women can be the consumers. I don't know [what] are drivers, because is the ivory itself or is it the swipe left product.

(Female 1.3.3)

So, women such as the wives of the poachers will try and deceit you and try and steer you off course to so that you don't find their husbands caught in the crime. When I was intercepting and undercover mission on I resale, there was three women in the room who are kind of the ones showing the ivory bringing it into the room showcasing it. So really just working together with criminal groups, because they're either wives or family members or siblings or parents or whatever, just because probably they received benefits from the crime from their counterparts and therefore, they see the benefit of engaging and sometimes.

(Female 1.3.2)

We deal a lot with men and trying to reform poachers and raise awareness with men and young boys in parks and scouts and so on. But the women are very much overlooked in that. So, the benefits for women in committing these crimes is quite high.

(Male 1.3.1)

Some voices expressed the physical and emotional harms associated with women's roles in wildlife trafficking, often framing risk against the ethic of caregiving, for example:

> Living in, you know, areas where there is trafficking, there is danger. So the kids are in danger. The kids are in danger of getting, you know, recruited for the travel. And there is this sense of insecurity that prevents people from bringing resources into these places. So it is a dangerous situation and women more than anyone else know that children need to be protected.
> (Female 1.3.3)

> Once arrested they are separated from their family and then they don't have access to their children. And then of course is almost because your neighbors know most of the time the people know close family members, the neighbors know what the person is involved in at the community level. So, there is that kind of stigma where people know that person X is actually in illegal activities.
> (Female 1.3.4)

> The fact that this is an illegal activity means that your job is risky. These people might decide to kill the person thinking that they will form the police. So, they're always living in fear. And again, I think when someone is involved in illegal activities, there's that social stigma associated to people do bad things. And that remains not just to that person but also to the whole family will be seeing the ante of these people are the model these people.
> (Female 1.5.1)

> …in areas where there's quite a militarized or violent kind of anti-poaching response in source areas, where you see a lot of men being killed… hundreds and hundreds of men being killed over the past five, six years. Now, these women suffer a lot because they're losing their husband, right? So there is a whole kind of range of losses from there from emotional, kind of an intangible loss, but there is also a very real material loss, in areas where people are quite dependent on agriculture, levels of education literacy among women are really low. And traditionally men are the income earners. Women are put in a very precarious and vulnerable position because they no longer have that kind of support network.
> (Male 1.4.5)

> The cost is not just financial; it is also to their own security.
> (Female 1.3.1)

Voices from the field inspire many questions about the role of women as beneficiaries and persons harmed. What is the surrounding political, environmental, economic, and health context for different beneficiaries or persons harmed? How much control does an individual have over access to benefits in public and private spaces? Are harms associated with wildlife trafficking linked to other forms of social conflict or exploitation, and if so, under what conditions? How do we test ideas that greater emancipation, labor force participation, and formal equality of women have fundamentally affected women's beneficial involvement?[3] Answers to these and other questions may help reduce harms associated with wildlife trafficking.

Notes

1 Agu, H.U., Andrew, C., and M. L. Gore. (2021). Mapping Terra Incognita: An Expert Elicitation Of Women's Roles in Wildlife Trafficking. Frontiers in Conservation Science. https://doi.org/10.3389/fcosc.2021.683979
2 The Michigan State University Human Subjects Protection Program approved the methods and analysis for the study (STUDY00003659) as exempt under 45 CFR 46.104(d) 2(ii).
3 Selmini, R. (2020). Women in Organized Crime. *Crime and Justice, 49*(1), 339–383.

11 Conclusion

Bridging Knowledge–Action Boundaries

Meredith L. Gore and Helen U. Agu

This edited volume was intentionally curated to help fill a lacuna in the knowledge base. The science on women and wildlife trafficking presented herein is novel, interdisciplinary, rigorous, and applied. These chapters help meet demand by management authorities and local practitioners who use evidence to inform action in response to wildlife trafficking. Chapters in this volume discuss inclusion, power, governance, and more; these considerations are relevant to the implementation, monitoring, and evaluation of wildlife trafficking programs and policies. Enhanced understanding about the roles of women in wildlife trafficking portends more effective and efficient policies and programs designed to conserve natural resources, promote gender equality, and women's empowerment. To be sure, some research at the wildlife trafficking knowledge-to-action boundary exists, but is underutilized (Milner-Gulland et al., 2018). This volume helps prove that clear research opportunities remain for new bridges to be formed across knowledge–action boundaries, and that multiple approaches exist to theoretically underlie such research.

Wildlife trade—legal and illegal—provides substantial benefits to humanity; harvested wild animals and plants are sought for food, medicine, ornaments, fuel, construction materials, and various other purposes linked to utility, recreation, and culture ('t Sas-Rolfes et al., 2019). Wildlife trade is *very* diverse, from live animals and plants to a vast range of wildlife products derived from them, such as food products, exotic leather goods, musical instruments made of wood, tourist souvenirs, and medicine. The levels of exploitation of some animals and plants are high and their trade, together with other factors, such as climate change and habitat destruction, is capable of considerably reducing their populations and even pushing some species to the verge of extinction. Illegal wildlife trade (IWT) is aided by the use of the Internet and new technologies helping offenders to evade capture, even while importing countries have attempted to crack down on the trade with increased funding and professionalized enforcement actions, such as through Interpol (McElwee, 2012). Women and men participate differently in all domains of IWT, from poaching through to consumption, and in policymaking. The costs and benefits of poaching, trafficking, and consuming illicit wildlife

DOI: 10.4324/9781003121831-15

can be different for men and women, as can be the costs and benefits of curbing or ending IWT. Access to resources, control over those resources, attitudes about them, and decisions about what resources to use can differ for men and women (Seager et al., 2021).

We see at least three categories of research that will help fill identified needs by the conservation and criminological communities, inspired by Bostrom et al. (2021) in the *Environmental Change and Human Security: Research Directions Report*. These categories of research can help build new bridges across knowledge–action boundaries associated with the roles of women in wildlife trafficking. First, what are the direct effects of wildlife trafficking on women? Wildlife trafficking occurs across different spatial and temporal scales, with differing effects on individual and groups of women. Quantifying and qualifying these effects in the different socio-environmental systems within which wildlife trafficking occurs will enable both inductive and deductive reasoning. Scientists can contribute holistic observations and identify new patterns and innovative explanations about women and wildlife trafficking in general as well as specific contexts (e.g., the involvement of Cameroon's Indigenous Baka Women as discussed in Chapter 8, the role of women in illicit supply chains as discussed in Chapter 6). Second, what are the relationships between global environmental changes and women's social stress? Here, science can explore the range of different activities that women engage in response to environmental changes such as shifting natural resource economies in Central African Republic, as discussed in Chapter 9. It is also important to explore how women's behavior may directly or indirectly drive environmental changes such as in Mexican natural resource institutions, as discussed in Chapter 4. Both at-risk and resilient situations (e.g., hotspots, coldspots) can be identified; the former can be more effectively managed, and the latter may be more systematically replicated in other conservation contexts. Third, what kind of questions might we anticipate from large-scale environmental surprises such as extreme climate events or smaller shocks that can have compounding effects such as harmful algal blooms? What are the local, regional, and global causes and consequences of environmental surprises at different scales for women and wildlife trafficking (e.g., wild meat consumption in urban centers vs. rural areas, as discussed in Chapter 6)?

Cook et al. (2013) opined that generating science capable of effectively informing decision-making (i.e., boundary science) requires that the production of knowledge be at least salient, credible, and legitimate. Without these characteristics, science is likely to be ignored by decision-makers. Or, new knowledge–action bridges will not be navigable. We recognize the diversity of challenges that can be associated with bridging the knowledge–action boundary, such as the proportion of women serving in decision-making roles as discussed in Chapter 4 or changes in regulations governing wildlife use as discussed in Chapter 10. Addressing these challenges may involve research that helps answer questions such as in what form will boundary science have the greatest impact (Bostrom et al., 2021)? Data about the roles of women and wildlife trafficking are collected using different methods, analyzed using comparative epistemologies and ontologies, and communicated via multiple mediums. Or, what could or should be the relative role of big data approaches, comparative case studies, and model development and simulations? Who should conduct the research? Science teams, individual investigators, academic networks, and practitioner partnerships are differentially positioned, funded, and resourced to advance production of knowledge. Boundary science may be challenging but advancing scientific understanding *and* contributing to decision-making can be a force multiplier for positive societal change.

Wildlife trafficking is a socio-environmental problem with far-reaching implications for biodiversity conservation and livelihood preservation. The roles of women in wildlife trafficking should no longer be understudied by scientists or underconsidered by practitioners. Women are half of the earth's population. They are part of the problem. They are part of the solution.

Literature Cited

Bostrom, A., Janetos, A., Clarens, A. F., Dabelko, G. D., Huybers, P. et al. (2021). Environmental Change and Human Security: Research Directions. Report prepared for the National Science Foundation. https://www.nsf.gov/ere/ereweb/reports/AC-ERE-Environmental-Security-Report-508.pdf.

Cook, C. N., Mascia, M. B., Schwartz, M. W., Possingham, H. P., & Fuller, R. A. (2013). Achieving Conservation Science that Bridges the Knowledge–Action Boundary. *Conservation Biology*, *27*(4), 669–678.

McElwee, P. (Ed.). (2012). *Gender and Sustainability: Lessons from Asia and Latin America*. University of Arizona Press.

Milner-Gulland, E.J., Cugnière, L., Hinsley, A., Phelps, J., 't Sas-Rolfes, M., & Verissimo, D. (2018). Evidence to Action: Research to address the illegal wildlife trade. *Briefing Note to Policy-Makers and Practitioners*. doi: 10.31235/osf.io/35ndz.

Seager, J., Parry-Jones, R., & Léger, T. (2021). Gender and Illegal Wildlife Trade: Overlooked and Underestimated. *Oryx*, *55*(5), 653–654.

't Sas-Rolfes, M., Challender, D. W., Hinsley, A., Veríssimo, D., & Milner-Gulland, E. J. (2019). Illegal Wildlife Trade: Scale, Processes, and Governance. *Annual Review of Environment and Resources*, *44*, 201–228.

Index

Note: **Bold** page numbers refer to tables; *italic* page numbers refer to figures and page numbers followed by "n" denote endnotes.

agency analysis 31
agents of change 30–31; illegal wildlife trade 30, 41; women 34–40; women in supply chains 32–33
Age of Enlightenment 13
agroforestry 111, 114–117
Agu, H.U. 27, 60, 69, 75, 79, 123, 161
Akashinga, Zimbabwe 38–39
All-Female Units 40
animal harm 155–156
Animal Welfare Act 153
anti-poaching solutions 18
Arzumanyan, L. 112
Association Communale Yobé-Sangha (ACYS) 134
Attorney General for Environmental Protection (PROFEPA) 47, 48–49; experiences, wildlife inspectors 53–54; key positions, wildlife departments **51–52,** 51–53; perceptions, wildlife inspectors 53–54
Ayling, J. 19

Bahuchet, S. 81
Baka community/people 107, 108, 109, 111, 116, 117, 131; sedentarization programs 110
beneficiaries, women's roles 61, 75, 79, *79,* 81, 161–164
Bilo community 131
Black Mambas, all-female anti-poaching unit 19, 39
border crossing, informal methods for 97
Bostrom, A.: *Environmental Change and Human Security: Research Directions Report* 166

Bride, I. 155
bushmeat trafficking 73, 76–78, 81–82; *see also* wild meat trafficking

Cáceres, B. 35
cafeterias 76–78
Cameroon: biodiversity 72–73; bioeconomy 72–73; channels of crime 73–75; conservation crimes 73; conservation-related behaviors in 115; Dja Biosphere Reserve (*see* Dja Biosphere Reserve (DBR)); indigenous people and conservation in 107–109; women 72–73
Carson, R.: "Silent Spring" 34
Cawthorn, D. M. 19
Center for International Forestry Research (CIFOR) 20
Central African Republic (CAR) 128; wildlife and conservation economies (*see* wildlife and conservation economies, southwestern CAR)
centrality, of gender and gendered relations 90
chivalrous model 5–6
community-based conservation projects 108
community-level perspective, rhino poaching 100–101
complementary spaces 126, 127
conflict theory of crime 2
Congo Basin forest 107; dynamic relationship of indigenous people 108
conservation: in Cameroon 107–109; and criminological communities 166; in Dzanga-Sangha Protected Area

128; and land use types 111; notions of inclusion in 106–107; perceived inclusion of indigenous Baka women in 114
conservation anthropology 139
conservation crime 73, 88, 93, 102
conservation criminology 148
conservation environment, defined as 111
conservation-related micro-landscapes 117
conservation scientist–activists 34–35
control points 77
Convention on International Trade in Endangered Species of Wild Fauna and Flora (CITES) 18, 143–145
Cook, C. N. 166
Cooney, R. 18
corruption 13–14
cost–benefit analysis (CBA) 76–77
crime 146; drug 154
criminality 3, 59, 63, 147
criminalization, of reptile keeping 154
criminal justice professionals 69
Criminal Justice, Wildlife Conservation and Animal Rights in the Anthropocene (CRIMEANTHROP) 148
critical criminology 88–89, 92, 93
Cronin, D. T. 17
cultural customs 100

Daspit, L.L. 131
data collection, on wildlife trade 147–149
Davies, P. A. 89
decision making 7, 75, 109, 112, 116–118, 123, 124, 139
defender, women's roles 27, 61, 62, 65, 69–71, 75, 79, 80, 82
demographic analyses 15–18
depression 65
direct abuse 155
direct contact 59
diversity, defined as 106–107
Dja Biosphere Reserve (DBR) 17, 109–110, *110*; analytical samples 112, **113**; mean inclusion scores for indigenous Baka women 114, **114**; measurement 111–112; protected areas-based landscapes 115, **115**; study population and sampling frame 110–111; *see also* Cameroon
domestic agricultural labor 96
drug crimes 154
Dzanga-Sangha Protected Area (DSPA) 127, 139; conservation in 128; gazettement of 129–131, *130*; *see also* wildlife and conservation economies, southwestern CAR

ecoguards 82
economic redistribution 56
electronic communications 61
elephant meat 138
Ellis, C. M. 81
emotional harms 163
employment opportunities 110; formal and informal 132
empowerment 6–8
engendered social justice 46
Environmental Change and Human Security: Research Directions Report (Bostrom) 166
environmental crime 2, 31–32
environmental harm 15, 146
environmental protection 54–56
Epanda, M. A. 17, 117
equity 41, 139
ethic of caregiving 163
ethnographic approach 127–128
European Union Wildlife Trade Regulations (EWTR) 144
exclusion 112; dynamics and consequences of 107; for indigenous people and local communities, history of 117
exploitation 117; levels of 165
expropriation, for indigenous people and local communities, history of 117

face-to-face communications 61
Fa, J. E. 116
family: and rhino poaching conflict 99–100; risk of exposure from offenders and defenders 62
female offenders, relationship with reptiles 151–152
feminist approaches: on poaching economies and militarized responses 87; for understanding poaching economies and responses 91–93
feminist criminologists 92
feminist criminology 87, 90, **90,** 91–92
feminist green criminology 92–93
feminist political ecological framework 87
feminist political ecology (FPE) 12–13, 87, 88–91, **90,** 91, 126; corruption 13–14; criminological issues 13–14; demographic analyses 15–18; green

criminology 14–15; history of 13; of poaching 93–101
feminist political ecology-based research 139
Feminist Political Ecology: Global Issues and Local Experiences (Rocheleau) 13
feminist scholarship 93
feminist social justice 47–48
feminist theory 45–46
Fimbel, C. 76
Food Safety Authority (FSA) 144, 150, 151, 156
foreign consumption 74, 77
1994 Forestry Law of the Republic of Cameroon 111
formal employment opportunities 132
Fossey, D. 34
Frank, Senator 35
Fraser, N. 47–48, 56
FRIPRO 148

gazettement: of Dzanga-Sangha Protected Area 129–131, *130*; impacts of 131–133, *132*; *see also* Central African Republic (CAR)
Gbaya ethnic group 131
gender 6, 45, 46, 62; centrality of 90; dynamics of 95
gender-as-culture 144
gender-disaggregated focus groups 17
gendered demographic analysis 16
gendered dimensions, wildlife trafficking 11–12
gendered norms: intersections between 93; structures of 92
gendered relations: centrality of 90; in rhino poaching economy 95–99, *98*; structures of 92
gender equity 37, 40
gender–power structures 92
gender role expectations 4
gender stereotypes 6
General Directorate of Wildlife (DGVS) 48
General Law of Ecological Balance and Environmental Protection (LGEEPA) 48
General Wildlife Law (LGVS) 48, 49
Gérard, A. (1965) 117
Givá, N. 102n1
Glajar, D. S. 18
Global Financial Integrity 19
global illegal trade, in reptiles 143–145
Goetz, A. M. 13–14

Goodall, J. 34
Gore, M. L. 17, 60, 75, 79
gray literature 19–20
Great Limpopo Transfrontier Conservation Area (GLTFCA) 98
green criminology 87, 89, 90, **90,** 149; data collection on wildlife trade 147–149; origins of 14–15; perspectives and hegemonic masculinity 145–147; relationship with wildlife trafficking 15
green criminology scholarship 44
green justice 54–56
Guatemala Maya Biosphere Reserve 108–109

Hardin, R. 134
harm 161; animal 155–156; environmental 15, 146; physical and emotional 163
hegemonic masculinity 145–147
Henry, B. C. 156
Herbig, J. 155
Higher Prosecuting Authorities in Norway 148
HIV/AIDS 62–63
Hoffman, L. C. 19
homo domesticus 46
homo sapiens 46
household-level perspective, rhino poaching 99–100
households: dynamics of 47; scaling impacts of rhino poaching and militarized responses to 99–101; wild meat consumption by 17; women, burdens on 32
Howson, C. 14
Hübschle, A. 14, 18
human actor model 30–31
human–environment interactions 90
human health, wildlife trafficking: lack of scientific attention 66; mental health 60; nexus of 59–60; physical 60; social and mental health dimensions 60; social health 60
human-only perspective 56
hunter-gatherers 107, 116
hunter-gatherer theory 31
hunting safaris 114–116
hunting zones 111

ICDPs *see* integrated conservation and development projects (ICDPs)
illegal wildlife economies 135

illegal wildlife markets 15
illegal wildlife products 1, 33
illegal wildlife trade (IWT) 30, 165, 166; women as agents of change 34–41; women in supply chains 32–33
inclusion: in conservation 106–107; opportunities for women 134
"Inclusion–Diversity–Equity" 106
income streams 133
indigenous Baka women (IBW) 108, 109, 118; inclusion in different workforces 117; interview respondents' perceived involvement of women in general and 115, **115**; mean inclusion scores for 114, *114*; perceived inclusion of 114
indigenous local knowledge systems (ILKs) 107
indigenous people and local communities (IPLCs) 106, 117, 118; in Cameroon 107–109; in Congo Basin forest 107; dynamic relationship of 108; history of exclusion and expropriation for 117
indigenous women, role of 108
indirect contact 59
industrialized logging 131
industrial logging 116
influencer 27, 61, 75, 79, 80, 82, 123–125; *see also* women's roles
informal employment opportunities 132
informal guardians 69
insider–outsider relationships 4–5
institutionalized harm 146
integrated conservation and development projects (ICDPs) 128, 133–135
Intergovernmental Science-Policy Platform on Biodiversity and Ecosystem Services (IPBES) 106
International Anti-Poaching Foundation (IAPF) 38
International Fund for Animal Welfare (IFAW) 36, 39
International Ranger Federation World Congress (2019) 37
International Union for the Conservation of Nature (IUCN) 73, 144
interspecies health 59
Ioveva-Baillon, K. 81
IPBES *see* Intergovernmental Science-Policy Platform on Biodiversity and Ecosystem Services (IPBES)
IPBES Global Assessment 106

IPLCs *see* indigenous people and local communities (IPLCs)
"Ivory Queen" 33
IWT *see* illegal wildlife trade (IWT)

Johns Hopkins University Gender Analysis Toolkit 6
Jost Robinson, C.A. 128–129
justice 146; engendered social justice 46; feminist social justice 47–48; green justice 54–56; social justice 46–56

Kahler, J.S. 17
key informant interviews 61, 63, 65, 76
knowledge–action boundaries 165–167
knowledge gaps 20
Kruger National Park 95, 97, 98; rhino poaching in 99

law enforcement authorities 69
leadership and management positions 36
legal wildlife economies 135
Limpopo National Park 97
lizards 150, 151
lobola 96, 99
local communities 38, 40, 59, 82, 95, 127, 129, 131, 134
logging: concessions in Dzanga-Sangha Protected Areas 131, *132*; industrial 116; industrialized 131
Lowassa, A. 17, 127
Lunstrum, E. 102n1
Lynch, M. J. 92

Maasai women 39–40
Maathai, W. 35
MacArthur Foundation 134
Macdonald, D.W. 116
mainstreaming 6–8
male-focused interventions 139
male offender 149, 150, 154
market chain analysis (MCA) 76–77
"market mamas" 19
market retailers 76
masculinity 74–75, 96, 152–155
Massé, F. 102n1
Mayrhofer, U. 112
Mbazza, P. 80
Mbete, R.A. 16
McElwee, P. 32
men: complementary spaces within society 127; gendered expectations for 96

mental health 60, 65
Merchant, C. 13
Metheny, N. 113
Mexico 44, 46–54
micro-environment 115; management of 116
micro-landscape economies 117
micro-landscapes type 114, 115, **115,** 116
migrant groups 131
migrant women 136–137
militarized responses: feminist approaches on 87; to women and households, scaling impacts of 99–101
Milliken T. 20
Mineka, S. 153
Ministry of Environment and Natural Resources (SEMARNAT) 47, 48–49; key positions, wildlife departments 51
Ministry of Wildlife and Forestry 80, 82
money-poor communities 73–74, **74**
Moreto, W. D. 18
motivations/intentions, offenders 27
Moyer, I. L. 5–6
Mozambique, rhino poaching in 94
Mozambique–South Africa borderlands 94, 96, *98*
multiple biodiversity conservation 108
multiple linear regression 112

national administration websites 76
National Geographic Magazine 36
National Institute of Cartography 76
net hunting 134, 138
Nigeria 59–66
non-Baka community 116
nongovernmental organizations (NGOs) 69, 108
Norway: animal harm 155–156; case for reptile trade 143–144; extrinsic token of masculinity 152–155; relationship of female offenders with reptiles 151–152; reptile keeping and trafficking in 149–151
Norwegian Centre for Research Data 148
Norwegian Environment Agency 147
Noss, A. J. 138

observer 27, 61, 75, 79, 80–83; *see also* women's roles
occupational socialization 4
offender 27, 61–63, 75, 79, 80, 148; convicted of drug crimes 150; female 151–152; interviews with 149; male 149, 150, 154; toughness 153; *see also* women's roles
Öhman, A. 153
Olivero, J. 107
One Health perspective 59
one-on-one harm 146
one-size-fits-all approach 12
"on-the-ground" practice, of conservation 90
operationalization, of patriarchy 92
Ordaz-Németh, I. 16
"orientation" strategy 31

Pateman, C. 46
patriarchal social organization 45–46
patriarchy 45–46, 91–93
Peace Corps 134
peer-reviewed literature 18–19
penal case file archives 148, 149, 156–157
perceived inclusion, of indigenous Baka women 114
Persaud, S. 18
persons harmed 161–164; *see also* women's roles
"pet reptiles" 156
physical harms 163
physical health 60, 61–63
Pires, S. F. 17, 18
Plumwood, V. 13
poaching 31–32; economies and responses, feminist approaches for understanding 91–93; economies, feminist approaches on 87; feminist political ecology of 90, **90,** 93–101; in Mozambican–South African borderlands 87; pillars of feminist political ecology *94*
poaching economy 93; advancing feminist perspectives on 101–102; long-term research on 94–95
policy drivers 35
Political Constitution of the United Mexican States [CPEUM] 48
political ecology 45–46, 88–90, **90;** *see also* feminist political ecology (FPE)
political economy, in rhino poaching economy 95–99, *98*
The Political Economy of Soil Erosion (Rocheleau) 13
Poulsen, J. R. 17, 116
poverty 72, 97–100
Price, R. 19

Priston, N. 17
protected areas 117; governance in Congo Basin 108; and hunting safaris 116; and hunting zones 111; to private logging concessions 108; wildlife conservation in 115
protected areas-based landscapes 115

Rainforest Foundation United Kingdom 108
Ramutsindela, M. 19
rangers 36–40
recognition 46–47
Recovering America's Wildlife Act (2021) 35
redistribution 46–47
reflexivity 137
Remis, M.J. 128–129
representation 46–48
reptile: global illegal trade 143–145; meaning of 152–155
reptile trafficking, Norway 144–145, 149–151; animal harm 155–156; extrinsic token of masculinity 152–155; relationship of female offenders with reptiles 151–152
Research Council of Norway 148
resistance-oriented dynamics 93
restaurants 76
Reuter, K. E. 17
rhino poaching: in Mozambique and South Africa 94; to women and households, scaling impacts of 99–101
rhino poaching conflict 99
rhino poaching economy 99; local gender relations, political economies, and women's participation in 95–99, 98
rhino poaching-related wealth 99–100
Robinson, J. G. (1999) 116
Rocheleau, D.: *The Political Economy of Soil Erosion* 13
Roosevelt, T. 35
rural consumption 73–74

sale points 76, 78
Sangha Trinational (STN) 129
Scarborough, K. 3, 4
Scientific Revolution 13
'secondhand animals' 156
security: anti-poaching unit with 19; conservation-induced 97; material 99, 100, 102; socio-cultural 100; socio-environmental 2

sex 6, 16
Shangaan culture 96
Sharma, S. 7
Shaw, J. 20
Shiva, V. 13
"Silent Spring" (Carson) 34
SINFOCAM 132
Slovenia Bois 131
snakes 150, 152, 153
snare hunting 137
social–ecological interactions 107
social health 60, 63–65
societal challenges 2
Société de BoisBayanga (SBB) 132
sociodemographic dimensions 12
socio-economic class 92
socio-economic/cultural benefits 27
socio-environmental conditions 124–125
Sollund, R. A. 18, 92
South Africa, rhino poaching in 94
South Africa–Viet Nam Rhino Horn Trade Nexus report 20
South, N. 14, 15
speciesism 146
Spira, C. 17
Stephenson, R. 113
Strobel, S. 82
subsistence agricultural labor 96
subsistence hunting 126

Tagg, N. 17
Taylor, G. 77
Team Lioness 39–40
temporary inclusion 112
Thai women 18, 19
3Rs (representation, recognition, and redistribution) tripartite model 46–47
Tieguhong, J. C. 116
traditional Asian medicine (TAM) 148
traditional Chinese medicine (TCM) 148
trafficking routes 77, 80–81
transfer-oriented models 108
transnational environmental crime 59

unbalanced interventions 127, 140
United Nations Convention Against Corruption (UNCAC) 44
United Nations Convention against Transnational Organized Crime (UNTOC) 44

United Nations Multidimensional Integrated Stabilization Mission in the Central African Republic (MINUSCA) 128
United Nations Office on Drugs and Crime (UNODC) 32, 143
unstable logging economy 133
urban consumption 74

Van Vliet, N. 80
victimization 89–90; deconstructing processes of 92
victims, women's roles 61, 75, 79, 80, 83
Vietnamese women 19, 20
voices from the field 27–29, 69–70, 123–125, 161–164

wage–labor employment 96, 100
Watt-Cloutier, S. 35
Weber, D. S. 18
Wilderness Act (1964) 35
wild game meat 133, 139–140
Wildlife Act, Uganda (2019) 2
wildlife and conservation economies, southwestern CAR: gazettement of DSPA 129–131, 130; historical ebbs and flows 128–129; ICDPs 133–135; impacts of gazettement 131–133, 132; inclusive interventions with fewer shadows 138–140; lead author story 137–138; legal and illegal wildlife economies 135; respondent story 135–137; separate income streams 133
wildlife conservation 117
Wildlife Conservation and Protection Act, Thailand (2019) 2
wildlife crime 11, 15, 146–147; Cameroon 74–75, 77–83; characterization of roles 79, 79–83; control points 77; economic importance to women 78–79; sale points 78; sex bias 27–28; trafficking routes 77, 80–81
wildlife economy: underexplored roles of women in 126–128; using ethnographic approach 127–128
wildlife inspectors 53–54
wildlife management, perceived inclusion of indigenous Baka women in 114
wildlife poaching 17, 18, 35, 89, 93, 102, 109
wildlife products 19, 54, 161, 162, 165
wildlife regulation initiatives: commissions 49–51, 50; promoters 49
wildlife trade 2–3, 6; data collection on 147–149; legal and illegal 165
wildlife trafficking: Cameroon 72–83; crime industries 11; demographic analyses 15–18; feminist political ecology 12–14; gray literature 19–20; green criminology 14–15; illegal wildlife trade 2–3; mental health implications 65; Mexico 44–56; Nigeria 59–66; peer-reviewed literature 18–19; physical health implications 61–63; and reptile trafficking 144–145; role of women as influencers in 123–125; social health implications 63–65; social justice 46–56; socio-environmental problem 167; sustainable conservation 6; women offending 28–29; women's roles 60–61
wild meat trafficking 16, 17, 19, 20, 33, 63–64, 115, **115**, 116, 138, 166; see also bushmeat trafficking
Wittig, T. 19
women: alternative livelihoods 82; as "assets" for conservation 30; as beneficiaries and persons harmed 161–164; Cameroonian women 72–73, 75; characterization of roles 79, 79–83; complementary spaces within society 127; conservation scientist–activists 34–35; demographic analyses 15–18; direct and indirect benefits from wildlife trafficking 161–164; economic importance of wildlife crime 78–79; empowerment 6–8; experience as foundation for status 101; gendered expectations for 96; and gendered relations function 92; influence individuals/groups of people 123; insider–outsider relationships 4–5; leadership and management positions 36; mainstreaming 8; mental health 65; motivations and intentions of 162; negative socio-economic impacts 89; offending proportion of 28–29; opportunities for inclusion 134; participation in rhino poaching economy 95–99, 98; physical health 61–63; policy drivers 35; private and public life roles 3–4; rangers 36–40; representation of 3–4; role

as influencers in wildlife trafficking 123–125; roles in wildlife trafficking 75; romantic connections 27; scaling impacts of rhino poaching and militarized responses to 99–101; social health 63–65; social roles 60–61; supply chain, IWT 32–33; as supporters 27; as unique approach 5–6; in wildlife economies, underexplored roles of 126–128; as wildlife trafficking defenders 69–70

World Wild Fund for Nature (WWF) 129; 2020 Living Planet Report 36
World Wildlife Crimes Report 19, 36
Wright, J. 17

Yang Feng Glan 33

zoonotic diseases 59–60, 62–63
Zwier, P. J. 18
Zwolinski, J. 116